COMFORTABLY
UNAWARE

COMFORTABLY
UNAWARE

What we choose to eat is
killing us and our planet

DR. RICHARD A. OPPENLANDER

BEAUFORT BOOKS | NEW YORK CITY

Beaufort Books

27 West 20th Street, Suite 1102

New York, New York 10011

(212) 727-0222

www.BeaufortBooks.com

ISBN - 978-082530686-0

LCCN - 2010939883

Cover Design & Typeset by Kristeen Wegner and Mike Naylor

Printed in the United States of America

📖 **BEAUFORT BOOKS**

To Lauren, Ricci, and Ty

for my perpetual inspiration to learn

And of course to Jill,

for continually showing me the path

toward some of the finer things in life...

love, belief, compassion,

and three beautiful children

Regarding Food:

"Every day each of us must make choices and then, ultimately, take responsibility for the comprehensive impact of those decisions. Therefore, it seems to be the inherent duty of everyone to make as informed a choice as possible. We should all be committed to understanding the reality and consequences of our diet, the footprint it makes on our environment, and seek food products that are in the best interest of all living things."

Richard A. Oppenlander, 1989

Comfortably Unaware **is about truth and about making a difference.**

CONTENTS

"If you care about our planet, and want to make a difference, this book is a must-read."
– Ellen DeGeneres's BOOKS ELLEN READS

"*Comfortably Unaware* explains, so clearly, how what we choose to eat has a direct impact on the health of Planet Earth: how modern agro-business and our thoughtless appetites are, quite literally, destroying the environment and the future of our children. I urge you to read it, to think about its message, discuss it with your friends — and start to change the world, one bite, one meal, one diet at a time."
– Jane Goodall, PhD, DBE, UN Messenger of Peace

"As vegan diets gain popularity across the country for a way to improve health and the welfare of animals, it's no secret that the environmental effects of this diet can have a positive effect on our planet. *Comfortably Unaware* helps readers take a closer look at just this — how to heal the planet by changing what's on your plate. A health and environmental advocate for over 30 years, Dr. Oppenlander has long been inspiring change with his informative message. May his message inspire you." – Neal Barnard, President, Physicians Committee for Responsible Medicine

INTRODUCTION

In the two-year period of time that was required to write this book and have it published, there has been an increase in attention—at least in some developed countries, such as the U.S.—given to the plight of our environment. Although known for decades by some, the effect on our planet of our choice of food as it involves livestock has finally made mainstream media headlines. Since 2007, a few authors have written books that begin to make the connection between the industrialization of foods and some—but certainly not all—of the effects on our planet. **Because of this, I feel _Comfortably Unaware_ will be read by two distinct audiences:**

One audience will have kept abreast of the news, read the books or heard the authors, and acquired a base of understanding about this connection, but their knowledge needs to reach another level of accuracy and comprehension, with unique perspectives, which _Comfortably Unaware_ will provide for them. Until now, they have heard only a story that is incomplete and, at times, inaccurate.

The second audience will be those who are vaguely (or not at all) aware of the connection of food choice and its effect on our planet—and there are many, many individuals in this category. For this group, every chapter will be enlightening.

If you are a reader in the first group, however, I would suggest that your focus as you read should be on having been afforded another level of awareness beyond what you have been exposed to by other authors or various media. For example, it is not just the excessive use of fossil fuel or contribution to global warming that should concern us with regard to raising livestock. Instead, we should be concerned about *all* the effects. It also is simply *not* sustainable for us to continue to eat animals, even if they are grass-fed. This is particularly true on a global food production scale. Additionally, we need to be aware of our oceans and the fishing industry as that affects global depletion. The impact of our food choices is not just a land-based issue. Our water supply is severely affected by our food choices as well, as is world hunger.

We are witnessing what could only be considered the sixth era of extinction on our planet because of an accelerated loss of biodiversity. The leading cause of this massive loss of plants, insects, and animals is our current choice of animal products as food. And regardless of whether the animals you eat are grass- or grain-fed, it is not sustainable for your own long-term health to eat them.

Most important, though, is the way in which this information is disseminated and ultimately used (or not used) to develop global strategies. *Comfortably Unaware* will provide you with information regarding *all* the areas of sustainability affected by eating animals and how various decision-making organizations are, in many instances, mismanaging it. Those of you who already feel somewhat enlightened may have recently been exposed to the information you'll find in the first few chapters, but your journey through the rest of this book will provide you with a new level of understanding, new perspectives, and new solutions. And, regardless of which group you may fit into as a reader, once you have fully read *Comfortably Unaware*, all the dots should be connected for you about this evolving and immensely crucial topic.

Once in a while, a story comes along that needs to be told. More infrequently, a story comes along that needs to be *heard*. This is one of those stories—except that this is not just a story. *It is very real*, and it affects all of us on earth. A number of individuals and large businesses and organizations would rather this story not be told; much of the public most likely does not want to hear it. *Comfortably Unaware*, however, reveals the truth about what we eat and what it is doing to us—and to the sustainability of our planet. It is not just another book about food. *Comfortably Unaware* is about being just that—comfortably unaware—about this truth and how we can make a change for the better.

The intent of *Comfortably Unaware* is to provide an unbiased informational base upon which readers can, at the very least, be afforded the opportunity to increase their awareness of food choice as it affects their lives and the life of our planet, each and every day. I wrote *Comfortably Unaware* to dispel widely held myths and offer a clearer and more truthful perspective about this suppressed subject. From this informational base, there is hope that proper decisions will be made to make a positive impact on the health of our world.

Food and the nutrients it contains are essential to our very existence. Food has been the nucleus around which social and cultural experiences occur. Food has been the reason for the success or failure of past civilizations, as well as for us as individuals today. Over the previous century, the food we've consumed in developed countries—and particularly in the United States—has become more industrialized and more commercialized. As such, the origins of our food have become less understood and less important, with little or no appreciation of the resources required to produce it. The current state of our food industry is forged by business and often political agendas, whereby the only measurement of success has been with economic standards. Unfortunately, this myopic and selfish view has created a food production system that overlooks public and environmental health. So strong has this industrial food system become that the realities of food origins and true cost of damages to our planet's health have been obscured and suppressed by a complex overlapping of large business interests and political, media, social, and cultural influences.

You and most other consumers make purchase decisions, including those involving food, based on one or a combination of factors: price, trusted recommendations or by association, convenience, taste familiarity, necessity, etc. Rarely does one choose and buy food items based on where, how, why, or from what that item was derived or its cost to our environment. Certainly, the true origin of what we eat—the path and story of how it arrived on your plate or in your mouth—should be known. This journey and the true cost in used resources and the effect on your health and that of our planet should be understood and taken into consideration with each food choice made. These food choice realities should also be placed back into the equation as a parameter of success or failure of our entire worldwide food production system.

Comfortably Unaware reveals facts and provides fresh perspectives by exploring how food choices affect our land, water, air, pollution, biodiversity, true sustainability, and our personal health. I structured *Comfortably Unaware* to resemble more of a symphony than a conventional nonfiction book—that is, the initial chapters serve as a prelude that establishes an appreciation and better understanding for hearing the crescendo—the final chapters that follow.

I would like to extend a special thank you to Dr. Jane Goodall for authoring "Harvest for Hope," which provides a global view of our current food choices with themes of sensitivity and hope. Other authors have paved the way and have written about food choices as they relate to our health and various disease states. To them, I am very grateful and acknowledge their commitment and accomplishments. Although there have been many, specifically I would like to thank Dr. Neal Barnard of the Physicians Committee for Responsible Medicine, and Dr. Dean Ornish of the Preventive Medicine Research Institute.

Over the years, I have lectured extensively on this topic to numerous audiences and have written many, many notes and articles. In fact, each chapter of *Comfortably Unaware* and many of the subtopics could have been developed into multiple books. The difficulty I found, then, was how to condense this enormously important subject into a book for the masses that would provide awareness, create intellectual stimulation, and effect positive change. I wrote *Comfortably Unaware* quite carefully, in a way that would present research without seeming too academic, that would relate difficult-to-believe facts without seeming too "in your face," and at the same time, offering new, challenging perspectives without appearing too theoretical or smug.

The first two chapters prepare the reader with a definition of *global depletion* as it relates to food, and they provide relevant facts and figures to serve as tools to help with appreciation for the rest of the book.

Chapters III through VII are devoted to each area of depletion—our air, rainforests, land, water and oceans, and pollution. This separates, in a clear format, each area of our earth that is becoming irreversibly depleted by our food choices.

Chapters VIII through XI are intellectual in nature but easy to read and compelling. They provide insight on how this crisis happened and how to solve it. These are the chapters that separate *Comfortably Unaware* from all other books, as they provide never-seen-before perspectives about our culture.

Numerous books have been written about various diets and food as it relates to our health. Many also are now available about global warming and climate change. *Comfortably Unaware* is the first to bring to light the much larger and more insidious issue of global depletion as it relates to food. I have not cut corners or suppressed topics to avoid exposing businesses, institutions, or individuals and I am not concerned whether or not it is a risky business move for me to write this book. I also have not withheld or modified information because it may be difficult for you, the reader, to accept it or because it may be culturally or socially overloading for you.

So, my agenda is clear: to provide you with complete truth and compel you to understand all the issues of this critical topic. It is my sincere hope that you become more aware of and sensitive to the ubiquitous effect of your food choices and that a positive difference can be made in your life and in the health of our planet and all its inhabitants.

DEFINING GLOBAL DEPLETION AND USE OF THE WORD "SUSTAINABLE"

Comfortably Unaware, first and foremost, is a book about sustainability—of our planet, our resources, and ourselves. At the same time, though, it is a book about food choice and responsibility, which are intertwined inextricably with the concept of achieving true sustainability (although "true" or "full" sustainability may be a difficult, if not improbable, state to achieve).

Global depletion is a term I have used over the years to describe the loss of our primary resources on earth, as well as loss of our own health due to our choice of a certain type of food. Therefore, global depletion essentially is about sustainability, but I feel we need to hear it from a different direction and with a more accurate view, through an unfiltered lens. Most of us have heard about the atrocities of factory farms, the issues with high-fructose corn syrup, and the industrialization or processing of foods with their contribution to obesity—all important topics. But these are simply small fragments of the picture. We need to move beyond that to understand the entire picture by connecting the dots and including our effect on all aspects of global depletion—topics such as loss of biodiversity, world hunger, sustainability of our own health, water scarcity, agricultural land-use inefficiencies and loss of our rainforests, pollution, and the state of our oceans and fish, as well as the effects on climate change. The largest contributing factor to all areas of global depletion is the raising and eating of more than 70 billion animals each year and the extracting of 1–2 trillion fish from our oceans annually. It's simply not sustainable.

Because of what can be viewed only as misuse or abuse of the word "sustainable," I am introducing and advocating use of the term **"relative"** sustainability. How "sustainable" is it to raise and eat ANY animal products in a RELATIVE sense, as compared to plant-based foods? How can we **best** use our resources? What foods will have the very **least** effect on our planet? Which foods **best** promote our own human health and wellness, and which are the **most** compassionate? Do we really *need* to slaughter another living thing in order for us to eat? Or, sadly, is it because we *want* to? In terms of sustainability, this is the way we must begin viewing things, in a **relative** sense, from this day forward.

Even as we deplete our natural resources, we add 230,000 new human mouths to feed each day. Water will become scarcer—predicted to be a 40 percent global shortage in just 18 years (over one billion people are already without adequate drinking water; two billion are without running water for cleaning and hygiene)—nearly one billion people are considered hungry, and six million children will starve to death this year. Nevertheless, of the 2.5 billion tons of grain harvested in 2011, half was fed to animals in the meat and dairy industries; 77 percent of all coarse grain went to livestock. Many of our planet's issues—dwindling resources, food security concerns, increased climate change, hunger and poverty, loss of biodiversity, pollution, declining human health and escalating health care costs, and the ravaging of oceanic ecosystems—can be eliminated or at least significantly minimized by a simple, collective change to a healthier, more peaceful plant-based food choice and thereby a more efficient, more compassionate food-production systems.

Almost everything we do, every decision we make every day, is based on our culture—what we've learned; what someone has told us is acceptable or necessary. After realizing by the end of the nineteenth century that bloodletting wasn't so healthy for us after all, we miraculously stopped, even though we had been doing it for more than 3,000 years. We are accepting culturally driven practices today, especially with food choices involving all animal products, that are much more unhealthy for our planet and for us than bloodletting—and by all counts, we don't have 3,000 years to get it right.

CHAPTER I

So It All Goes

Defining global depletion as it relates to food
(where global warming fits)

"Discontent is the first necessity of progress."
—Thomas Edison

STOP. TAKE A STEP BACK AND ASK

yourself, "Where did this food I am about to eat really come from? How much water, land, and other resources did it take to get it from point A to point B? Why am I eating it?" Have you ever asked yourself that? Of course you haven't—and why would you? Where your food comes from has to be the most "out of sight, out of mind" process that exists in our culture today; it's obscured by many layers of cultural, political, and educational untruths and misperceptions. This is particularly true as it relates to our use of animals in the meat, dairy, and fishing industries. And yet, that very same subject—the origin of your food—is the cause of billions of unnecessary dollars spent annually on certain aspects of health care and loss of productivity. Most important, it is the major contributing force in *global depletion*—the eventual loss of our drinking water, air quality, land, biodiversity, and other resources.

Is global warming an issue with you? Whether it is or is not, please note that our current food choices detrimentally affect climate change and global warming more so than do all the cars, planes, trucks, buses, and trains used worldwide.[1] That might be shocking and difficult to accept, but it is important to know—and it is true. Also, while we certainly should be concerned about global warming, it is just one aspect of the much larger issue of global depletion. If you really want to reduce your negative impact on our earth, it is not so much a point of adding insulation to your house, for example, as it is what you elect to eat. While it is clear that we must be aware of global warming, it frankly does not matter how many light bulbs we switch out if we run out of water to drink. Nor does it matter what type of hybrid car we drive if we run out of clean air to breathe.

So while it seems that our collective attention has been on global warming, it is only a small fragment of the more complex picture of what we are doing to ourselves and to our planet, as it is one component of the bigger picture of global depletion.

What exactly is global depletion? It is the loss of our renewable and nonrenewable resources on earth. At this point, we may need to redefine "renewable" as it relates to our resources. For instance, water is generally viewed as "renewable" and yet some of the water that is used daily on our planet comes from sources that take thousands of years to create. Similarly, the trees and the ecosystems of living things that are dependent on those trees that are destroyed in the Amazon rainforest required hundreds of years to develop. How "renewable" are these? We should use the term "nonrenewable" for any of those resources that, if destroyed, would most likely not be seen again in our lifetime. This also should apply to animals, such as those found in rainforests and rangelands, whales, and all other marine life.

It is these life-sustaining resources that are being used or destroyed at a rate such that replacement or restoration is impossible for hundreds, if not thousands, of years—if ever. Water, land, air, and wildlife ecosystems are most affected, and while many industries are to blame, our food has had the largest single negative impact on our environment. Every day, individuals and various industries use our planet's natural resources. Land is used for housing, transportation, waste management, and agriculture. Our clean water supply is used for drinking, waste removal and cleaning, and agriculture. Fossil fuels are also in demand for personal use, as well as by a number of businesses, including agriculture. While most uses of our resources can certainly be scrutinized, modified, and even reduced, it is startling to know that the sectors that use and deplete most of our resources are the meat, dairy, and fishing industries. The choices of food that we all make directly impact the use and subsequent depletion of our planet's resources.

Because it is the current "buzz," let's begin with a brief overview of global warming, or "climate change." Experts and organizations have filtered much of our understanding of this subject, but essentially, global warming is principally caused by an increase in greenhouse gases. This assumes, of course, that the earth's relationship with the sun has not changed, and the energy derived from the sun remains constant. Because our attention has been primarily on the *production* of these greenhouse gases, it also assumes that there is nothing else on our planet that affects these gases in terms of absorption, which creates more or less of an effect on our climate. But there actually is something else—our forests. This is particularly true of our rainforests, which absorb vast amounts of carbon dioxide from the atmosphere, while producing and exchanging it for oxygen.

Although water vapor and ozone (O_3) are considered greenhouse gases, it is widely understood that the increase by human activities in the other greenhouse gases—carbon dioxide (CO_2), methane (CH_4), and nitrous oxide (N_2O)—have had the most influence of any factor on global warming. While atmospheric concentrations of CO_2 have risen by 35 percent from the preindustrial year 1750 to 2006, those of methane have risen 145 percent, primarily due to the rise of the livestock/meat/dairy industry.[2]

Much emphasis has been placed on reducing our dependence on fossil fuels and subsequent CO_2 production and what we, as individuals, can do to help reduce this trend. Former vice president Al Gore's book, *An Inconvenient Truth*, certainly has helped increase awareness about global warming and has provided a sense of authenticity to its existence—and this was a wonderful thing. Unfortunately, Mr. Gore did not tell the real truth, which set forth a public misconception that global warming is the primary concern with regard to our effect on the planet, and that our excessive production of carbon dioxide is the principal factor in global warming. Overall, Mr. Gore told only that part of the global warming story that was the easiest for him to explain—

essentially, that which was the most "convenient" for him. His story and proposed solution is the least controversial route and, despite the title of his book, actually is the easiest for all of us to accept and to act on. Ironically, he effectively chose the path of least resistance. He emphasized that the culprit of global warming is carbon dioxide. After all, it comprises 72 percent of emitted greenhouse gases, and we humans produce most of it by the electricity we use and the cars we drive—how easy to deal with it. Simply reduce electricity and drive less often.

So what exactly did Al Gore fail to mention about global warming? That carbon dioxide is not the cause of global warming. It is, however, one component, and the most convenient to talk about, and the easiest topic for which to draw up a list to help solve it. While it's important to be aware of and to minimize CO_2 emissions from cars and industry, the single most devastating factor that affects global warming and our environment is caused by what you eat.

Another profound example of failure to adequately mention or address the effects of food choice and climate change can be seen with the management of the Kyoto Protocol, adopted in 1997 by numerous countries. Although the intent to reduce greenhouse gas emissions is to be applauded, proper attention was not given to the effects of food choice, specifically from the meat, dairy, and fishing industries. The Kyoto Protocol developed from the United Nations Framework Convention on Climate Change (UNFCCC), which was initially an international treaty joined by a number of countries to begin considerations regarding global warming—or at least to mitigate the human influencing factors. The Protocol ensued and was approved by many nations and now has legally binding and stronger measures. This

Protocol sets forth targets for reducing emissions of greenhouse gases, and plans for "emissions trading" and "joint implementation" as incentives for participating countries. In its Methods and Science section, the Protocol does address deforestation and land use issues but never directly mentions that the largest contributing force for combined deforestation and land use concerns are the meat, dairy, and fishing industries, which are driven by our collective choice of foods to eat. This obviously is shocking when you think about their blatant omission of any suggested action to combat one of the largest causes of global warming. While a global community of governments and scientists have begun a more concerted effort to reduce our impact on climate change through treaties such as Kyoto, it is clear that much more of their focus should be toward solving our problem of food choices and production methods. Otherwise, any effort to make a positive impact on climate change will be futile.

Both methane and nitrous oxide are much more powerful than carbon dioxide as greenhouse gases. Methane has *twenty-three times* the global warming effect potential as carbon dioxide. Approximately 40 percent of all methane produced by human activities is from livestock and their flatulence and manure, to the point where atmospheric concentrations have risen 145 percent in just the last fifteen years. Nitrous oxide is *310 times* more powerful as a greenhouse gas than carbon dioxide. Our livestock industry generates 65 percent of all human-related nitrous oxide.[3]

These statistics should provide insight into a more complete picture of greenhouse gases and global warming, but it is not what we should most be concerned with. First, we must be healthy, and our planet needs to be healthy in order for us to survive. For our planet to be healthy, we need to be concerned about our water, land, air, and living ecosystems. Greenhouse gases and their effect on global warming is only one aspect of the complete picture, and CO_2 is only one greenhouse gas. It is what we eat and the choices we make in our diet, *not* the car we drive, that affects our supply of water, land, and air and will affect our success or failure on our planet.

Why is this the first time you have heard any of this? Many times we hear only what others want us to hear. This is particularly true when facts have been discovered regarding a sensitive topic that is being sheltered for one reason or another by large business or our government. For instance, there are a number of documented instances where, during the Vietnam War, our government and the media only allowed certain stories and images to make their way to the public. This was in order to minimize an already unfavorable public opinion toward the war. To a lesser extent but similarly, this has been the case with the war in Iraq. Occasionally, there will be an aviation report surfacing from NASA or the FAA that begins to divulge the reality of how congested our airways really are, providing numbers per day of near collisions. These reports quickly disappear from the headlines, as it is decided that they are most likely too much information for a newly concerned public. However, the most profound example of withholding or obscuring the truth is with the food we eat—the truth of what it really is, the reality of where it comes from, and what it does to us and to our environment.

You may be aware of global warming and consider yourself part of the "green" or "sustainable" movement. Those terms, however, are now almost overused and, at times, misused. For many, it is becoming the "cool" or socially or even politically correct thing to do—and in one sense, that is a good thing. However, it not so cool to think of yourself as "green" or sustainable because you recycle or you change to energy-efficient light bulbs when you still eat animal products that have a much more profoundly negative impact on our environment than all the bulbs you just switched out. Consider going one step further and actually *becoming* environmentally conscious. Instead of just *saying* you are "sustainable," do the right thing for yourself and for the planet and eat only plant-based foods. Then you really will be sustainable.

CHAPTER II

For the Unaware

Pertinent facts and figures

"Real knowledge is to know the extent of one's ignorance."
—Confucius

IT IS DIFFICULT TO DEVELOP A BET-

ter understanding of a particular topic when there is pre-existing confusion and imprecise use of certain words regarding that topic. With food choices, improper connotations abound. With this in mind, let's begin by correctly defining some terms and concepts.

Food: that which is consumed to support life.

Plants and animals are not "food" unless you choose to eat them. Plants are living structures with chlorophyll-containing cells, capable of taking carbon dioxide out of the air in exchange for producing oxygen. Plants have no blood, organs, brains, or nervous systems. Animals are living organisms that have saturated fat and cholesterol associated with all their cells and tissue. All animals have blood, organs, brains, and nervous systems, feelings, and emotions.

Animals are, in fact, animals—not meat. "Meat" is a term contrived and used by humans to obscure the reality of what they choose to put in their mouths. Animals (cows, pigs, sheep, fish, birds such as chickens and turkeys, etc.) are also not protein. Again, animals are *animals*. Protein, on the other hand, is a nutrient and can be found in many living things, including plants, and can even be found in lettuce. Fats and carbohydrates are also nutrients. Some animal parts that are eaten have more fat content than protein, and yet I never hear people say as they eat meat, "I need to get my *fat* today." Not all types or forms of protein, fat, and carbohydrates are needed by the body, nor are they healthy for us. All essential protein (and amino acids), fats, and carbohydrates ("essential" meaning those that are needed to sustain life and that we cannot produce ourselves) can be derived from plants. All animals and animal products, if eaten, contain many non-essential and, in fact, unhealthy substances, such as cholesterol, saturated fats, high levels of methionine-containing (sulfur-type) proteins, polycyclic aromatic hydrocarbons (PAHs), heterocyclic amines, hormones, and some pesticides, herbicides, and heavy metals.[4] Additionally, animals and animal products that are eaten contain no fiber, no appreciable vitamin value, and phytonutrients, which boost our immune response and other systems. Plants, on the other hand, contain hundreds of these very important substances. Although some researchers have known much of this information for at least the past fifty years, organizations such as the American Dietetic Association, the American Cancer Society, and others finally support these facts.

How does this relate to global depletion? In the United States, as well as in other developed countries, there is an unnecessary and unhealthy dietary dependence on animals and animal products. We collectively raise, feed, water, kill, and eat over 70 billion animals each year for food.[5] That number again: *70 billion*, which is ten times as many people as we have on the entire earth. In doing so, these animals use and deplete our renewable and non-renewable resources—they use food, water, land, air, and fossil fuels or other energy sources that could or should be used for us. We have developed a complex system of producing more and more animals that use more and more of our resources, while leaving a massive amount of waste, pollution, and adverse climate change in their wake. And it repeats itself, year after year, in alarmingly increasing volume and intensity—meat and dairy production is expected to double in the next ten years, and fishing production even more so. This system also has become complicated in that it is now heavily intertwined with our culture, politics, economics, and the suppression of the reality of its effect on our planet. The following are just some of many facts and figures with regard to global depletion:

- Global warming ("climate change") is caused by the production of methane, nitrous oxide, and carbon dioxide, not by carbon dioxide alone.
- Global warming is also caused by destroying trees and vegetation that regulate carbon dioxide and oxygen.
- Global warming is just one small component of global depletion.

- Methane is 23 times as powerful as carbon dioxide, and nitrous oxide is 310 times as powerful as carbon dioxide for their global warming potential.
- Forty percent of methane and 65 percent of nitrous oxide produced by all human activities are from livestock.
- Rainforests are the lungs of our planet, producing over 20 percent of the earth's oxygen.[6]
- Rainforests take millions of tons of CO_2 out of our atmosphere and store it in soil.
- Seventy percent of our rainforests have been slashed and burned in order to raise livestock.[7]
- Fifty-five percent of our fresh water is being given to livestock.[8]
- Over 70 percent of the grain in the United States is fed to livestock.[9]
- It takes 10 to 20 gallons water to produce one pound of vegetables, fruit, soybeans, or grain.[10]
- It takes over 5,000 gallons of water to produce one pound of meat.[11]
- One pound of vegetables, fruit, soybeans, or grain is healthier for you to eat than one pound of meat.
- During every one second of time, just in the United States alone, 89,000 pounds of excrement is produced by the chickens, turkeys, pigs, sheep, goats, and cows raised and killed for us to eat.[12]
- One acre of land, if used for vegetables, grain, and/or legumes, produces ten to fifteen times more protein than if devoted to meat production.[13]
- Over 30 percent of all usable total land mass on earth is used by livestock.[14]

- Over 80 percent of all arable (agricultural) land in the United States is used for or by livestock.[15]
- Six million children in the world will die from starvation this year.[16]
- 1.1 billion people in the world are considered malnourished or suffering from hunger.[17]

Although this information may seem rather stark, it is only because, for a number of reasons, we have been "comfortably unaware."

CHAPTER III

It's in the Air

*Depletion as it affects oxygen and the
quality of the air that we breathe*

"Don't blow it; good planets are hard to find."
—Unknown

THE AIR THAT WE BREATHE AND

our atmosphere, in general, are fundamentally necessary for life
on earth. It should not be taken for granted or abused, yet we
currently are doing both. At any point in time during the day,
are you aware of the air you are breathing or appreciative of the
oxygen it supplies? Probably not. We breathe, on average, fifteen
breaths per minute, 900 per hour, and 21,000 breaths each day.
With every breath, we need fresh air and the right amount and
ratio of oxygen. Our atmosphere serves many purposes, such as

regulating temperature and carbon, nitrogen, oxygen cycles, and protecting us from injurious radiation. These processes are complex and fragile, and human activities affect these in negative ways, such as climate change and pollution.

Some human activities have a larger negative impact than others, with livestock clearly having one of the greatest roles. Nearly every step in raising the billions of animals for food each year creates some form of depletion or degradation of our air. There are three primary ways this occurs:

- Through greenhouse gas emissions
- By pollution
- By changing water cycle processes and oxygen-carbon respiration through vegetation loss

At least two separate studies of the Antarctic Dome Ice Core confirm that human activities have resulted in our escalating present-day concentrations of carbon dioxide and methane, and that they are the highest that these greenhouse gases have been in the last 650,000 years of earth history.[18] Methane concentrations have increased by about 150 percent since 1800.[19]

The livestock sector is responsible for nearly 20 percent of all greenhouse gas emissions, measured in CO_2 equivalent.[20] Global transportation, on the other hand, accounts for 13 percent of all greenhouse gas emissions. Put another way, what you currently decide to eat every day creates more global warming than all the cars, planes, trains, buses, and trucks in the world combined.[21] The reason I say "currently decide to eat" is because through your food choices, you are ultimately responsible for the demand for meat and raising the 70 billion animals each year that causes this large part of the global warming issue and the much larger global depletion problem. If you simply stop the demand by choosing a plant-based diet, and the largest component of global warming and depletion will go away.

Animals raised for food emit large amounts of greenhouse gases in different ways. Directly, livestock emit carbon dioxide from the respiratory process, and we have 60 billion more animals on earth than humans, all of which breathe in oxygen and breathe out carbon dioxide. Additionally, all livestock emit methane, nitrous oxide, ammonia, and carbon as part of their digestive process, in the form of flatulence, manure, and urine. In the United States alone, livestock produce 89,000 pounds of excrement every second—that's 130 times as much as the entire human population of the country.[22]

The 2010 Agriculture and Air Quality Symposium, sponsored by the Institute for Livestock and the Environment, identified significant air pollutants caused by raising animals for food and called for establishing methods of control and reduction. In addition to greenhouse gases, other air pollutants that cause concern are volatile organic compounds that are precursors to ozone, hydrogen sulfide, many types of particulate matter, and ammonia and odor.

Indirectly, livestock adversely affect the carbon balance of the land used for feed crops and pasture, as well as with the massive amounts of fossil fuel used in the production process, including feed production, processing, multiple levels of transport, and marketing of livestock products. Looking at just the agriculture sector, livestock constitute 80 percent of all emissions.[23] While this figure reflects a global issue, there is variation on the local level, with some countries such as Brazil contributing 60 percent of its total greenhouse gas emissions from livestock, due to the very large cattle operations and corresponding destruction of rainforest. Livestock emit 10 percent of all CO_2 and 40 percent of all methane (twenty-three times the global warming potential of CO_2), 65 percent of nitrous oxide (310 times the global warming potential of CO_2) and two-thirds of all ammonia emissions, which cause acid rain and acidification of our ecosystems.[24] This makes your choice to eat meat one of the largest sectors for CO_2 emissions and the single largest contributor of methane and nitrous oxide and ammonia. Producing one calorie of animal protein requires more than ten times as much fossil fuel input and produces more than ten times as much CO_2 as does one calorie of plant protein.[25] And producing any meat from animals creates the production of methane, nitrous oxide, and ammonia, while producing food from plants creates none.

The enormous amount of land and forests cleared for livestock and our demand to eat meat creates other losses in numerous ways. When livestock destroy vegetation, either through clearing forests and land for feed-crop growth, or directly by the livestock themselves, it disrupts the normal water cycle processes in that area. This destruction of natural ecosystems creates a significant impact on climate change as follows:

- Carbon dioxide that has been stored by plants is actually released back into the air.
- Oxygen is no longer created and released by all the plants that have been destroyed.
- Carbon dioxide is no longer taken out of the air by the plants that have been destroyed.
- The newly deforested land becomes vulnerable to erosion

and eventual desertification, both of which are occurring at an alarming rate.

All food comes from somewhere and requires some degree of effort to produce, process, and transport. With meat, dairy, or fish products, we affect vastly more resources than we would by consuming plant-based foods—resources such as our air and—although it's not readily seen—the quality of the air and atmosphere around us is becoming depleted with each and every bite of food made from animal products.

It is quite clear that the ability of our planet to produce oxygen from our forests and oceans is being compromised, and the comprehensive aspect of our atmosphere is changing in an unhealthy manner. Both are not easily reversed—at least, perhaps, not in our lifetime.

CHAPTER IV

Rainforests

Depleting the lungs of our planet

"When one tugs at a single thing in nature,
he finds it attached to the rest of the world."
—John Muir

WHERE DO I START HERE? TALK

about a comfortable state of being unaware. But why should you have any interest in or be concerned about some trees that grow somewhere else in the world? Because while you order your steak or burger for dinner, another acre of rainforest—and all the life it contains—is destroyed. Sounds ridiculous, doesn't it? After all, you are just ordering food in a restaurant because you are hungry, and you have other things to think about—your job, the economy, clothes, your car, your next vacation. Where does the rainforest come in?

Well, first of all, rainforests exist, even though you won't drive past one on your way home, and you most likely have not invested in the rainforest in any of your retirement funds. Let's go back to that burger you ate at lunch—or any other meat that you have eaten over the course of your lifetime. Where did it come from?

Over 70 percent of the Amazon rainforest has been destroyed—lost forever—due to cattle ranching. The United States is the single largest consumer of Central and South American beef.[26] A startling 95 percent of Brazil's Atlantic coast rainforest has been slashed and burned, the vast majority of it to raise cattle.[27] Although it is not commonly known, approximately 34 million acres of rainforest on earth are lost each year.[28]

Consider this: When fires occur in California, it is broadcast on the news. During October 2007, for example, when approximately 190,000 acres in California were lost, there was seemingly non-stop news coverage. That same year, over *30 million* acres were lost in the rainforests, with no news coverage whatsoever. Is one circumstance really less devastating than the other? In fact, over 30 million acres of rainforests per year have been lost *every year* since the 1970s. Although some of this rainforest land is logged, most is slashed and burned, then used to either raise cattle or to raise crops to feed to cattle. As much as 80 percent of all global rainforest loss is turned into grazing for cattle or crops for livestock, and the process is extremely land-intensive. It requires fifty-five square feet of rainforest to produce just one quarter-pound burger. The crops grown on cleared rainforest are used to feed not only cattle but also chickens, turkeys, and pigs. In one crop season alone, 2004–2005, more than 2.9 million acres of rainforest were destroyed, primarily to grow crops for chickens used by Kentucky Fried Chicken.[29]

Another crop that is grown is soy, but not for direct human use. Soy used directly for veggie burgers, tofu, and soy milk in America is almost exclusively grown in the United States, but 80 percent of the entire world's soy crop is produced and fed to farmed animals. Most of this soy is now grown on rainforest-cleared land.[30]

You may say, "Big deal—what good are rainforests? They're just some trees somewhere else in the world that I will never see. I would rather have my meat." First and foremost, the rainforests produce more than 20 percent of the world's oxygen supply. They provide an environmentally essential task of continuously recycling the air and pulling CO_2 out of the atmosphere, while putting O_2 back into it. So with every acre of rainforest lost to support the meat industry, the earth loses part of its lungs and the ability to breathe and produce a fresh supply of oxygen—fourteen tons of oxygen per acre per year—while taking out tons of global-warming CO_2.[31]

Fifty years ago, 15 percent of our planet was composed of rainforest; today, this has been reduced to less than 2 percent. Despite this loss, almost 50 percent of all types of living things (equaling five million species of plants, animals, and insects) reside in rainforests. Although numerous species have yet to be discovered, scientists estimate that at least one hundred species per day are lost when the forest is cut down. In Brazil's Atlantic coast rainforest, of which less than 5 percent remains today, 70 percent of its plants and twenty primate species are endemic (they are found nowhere else in the world). The enormous biodiversity of the rainforest implores respect and the need to preserve it, not to destroy it. Less than 1 percent of its millions of species have ever been studied by scientists. One pond in Brazil can sustain a greater variety of fish than are found in all of Europe's rivers. Just a two-acre area of rainforest may contain over 750 types of trees (more than the total tree diversity of North America) and 1,500 species of plants. The demand for meat and the subsequent loss of rainforests has been responsible for the disappearance of over ninety different Amazonian tribes.[32]

Descendants of the Amerindians have lived in the rainforest for some 20,000 years, with traditions that have allowed them to exist in harmony with the forest without destroying it.[33] Today, with massive destruction due to our demand for meat, there are fewer than 250,000 native people living in the Amazon forests, where there were once more than six million.[34] Many scientists believe there are as many as fifty different indigenous groups still living in the depths of the forests that have never had contact with the outside world. As these ancient forests are cleared to make room for cattle or feed crops for livestock, habitat is lost to tribes and their sustainable way of traditional life. Once the forests die, so do the Amazonian people. These tribes, with their medicine men, or shamans, have a wealth of knowledge, particularly the medicinal properties of the thousands of plants found in the rainforests—knowledge that will die with them. A single tribe may use more than two hundred species of plants for medicinal purposes alone.[35] Because of our demand for meat, there has been needless destruction of ancient rainforests. With this, there has been the unfortunate loss of indigenous tribes and their shamans. And when a medicine man dies, the world loses thousands of years of knowledge that is irreplaceable ... but at least you get a burger out of it.

Perhaps destroying living things and creating extinction of species still does not hit home, so let's look at another aspect: over two thousand plants have been discovered in the rainforests that have anti-cancer properties. All botanists agree that not only are there many thousands more to be discovered but that many species are lost daily as the forests are destroyed to provide meat for the world. Medicines derived from the rainforest include:

- curare (muscle relaxant used in surgery)
- diosgenin (birth control, arthritis, asthma)
- ouabain (heart medicine)
- quinine (malaria, pneumonia)
- emetine (bronchitis, dysentery)
- vincristine (Hodgkin's disease, leukemia, and other cancers)

Vincristine is extracted from the rainforest plant periwinkle and is one of the most powerful anti-cancer drugs. It dramatically increases the survival rate for acute childhood leukemia. One-fourth of all prescription drugs and over 70 percent of all cancer treatment medications originate from the rainforest.[36] It makes no sense whatsoever to destroy rainforests and all the life they contain to raise cattle or grow crops to feed animals, when plants can be grown elsewhere for us to use directly as food.

Plants in general and rainforests in particular serve as natural sinks for atmospheric carbon dioxide, sequestering and storing it in vegetation and the soil. The destruction of vegetation leads to carbon release, loss of the ability to remove carbon from the atmosphere, and loss of the ability to create oxygen, which negatively impacts water cycles and reduces infiltration capacity and storage of the soil and increases runoff.

When millions of acres of forest, especially rainforests, are cut down, we lose in many ways:

- We lose the ability to filter harmful levels of carbon dioxide out of the air.
- We lose millions of tons of vital oxygen released into the air we breathe.
- We lose because of the millions of tons of carbon dioxide that is released into the atmosphere by the burning of trees.
- We lose by changing the soil from its absorbing moisture and detoxifying oxygen to its being deforested, erosive and, on average, allowing for only eight years of grazing and growing crops for cattle before it has become depleted.
- We lose entire ecosystems of plants and animals—one mature rainforest tree can support three hundred to five hundred different types of plants and hundreds of species of animals

Rainforests are cleared, slashed, and burned for the timber value and then for farming and ranching operations to support the meat requirements of the world. Although local operators and businesses have some responsibility, much of the rainforest loss to support the livestock industry is accomplished by world corporate giants, such as Texaco, Unocal, Georgia Pacific, Cargill, and Mitsubishi Corporation. Regardless, the real blame for the depletion of our vital rainforests lies with the consumer who creates the demand for animal products.

CHAPTER V

Whose Land Is It Anyway?

Global depletion of our land

"We will be known forever by the tracks we leave."
—Native American proverb

RAISING ANIMALS FOR PEOPLE TO

kill and eat requires massive amounts of land, water, food, and energy. With upwards of 70 billion animals raised each year in the livestock industry, enormous amounts of land are needed for their living space and grazing, and to grow crops to feed them. In the United States, nearly 80 percent of all land used for agriculture is used in some way to support the animals we eat.[37] That is half the entire land mass of our country. More than 260 million acres of U.S. forest have been cleared just to grow grain to feed livestock.[38] Livestock occupy 30 percent of all land mass on earth, and another 33 percent of all agricultural land is used to produce genetically modified organism (GMO) crops to feed these animals.[39] Hence, a solid portion of all the land mass on earth is used in some way to produce animals that we then kill and eat. It has been said that an alien ecologist observing earth might conclude that cattle is the dominant animal species in our biosphere.

Let's put it into perspective: on any given acre of land we can grow twelve to twenty times the amount in pounds of edible vegetables, fruit, and grain as in pounds of edible animal products.[40] We are essentially using twenty times the amount of land and crops and hundreds of times the water, as well as polluting our waterways and air and destroying rainforests, to produce animals to kill and eat ... which is unhealthier than eating the plant products we could have produced.

Using most of our land to support livestock is just one issue in the depletion of land caused by our choices in food. Other issues in this same category include land destroyed by overgrazing, biodiversity loss, and food depletion, all of which are related to livestock. Cattle and other livestock not only currently use a massive amount of our land, but destruction also occurs with overgrazing the land, which then causes erosion, loss of topsoil, and desertification—the land ultimately becomes a barren desert.

It is estimated that 700 million acres of U.S. rangeland have been degraded by overgrazing of livestock.[11] What does this mean? We have lost seven billion tons, or one-third, of the topsoil in our country. What needs to be understood, though, is that six billion of the seven billion tons of eroded soil is directly attributed to grazing and unsustainable methods of producing crops for cattle and other livestock.[12] This problem can also be seen in other countries, as approximately 20 percent of the pastures and rangelands in the world have been degraded as a result of livestock. Lost topsoil is particularly alarming in dry-area rangelands, where 73 percent has been eroded.[43] Although some erosion of topsoil can occur by natural means, the vast majority of measured loss is because land is being changed from supporting an evolved plant and wildlife ecosystem to a quite unnatural decimation for and by livestock. Worldwide, there has been quite rapid converting of natural habitat to pastures and cropland to feed animals, with more land converted between 1950 and 1980 than in the previous 150 years.[44] While expansion of pastures and cropland for animals is increasing in North Africa, Asia, and the Caribbean, it is greatest in Latin America and sub-Saharan Africa, at the expense of forest cover.[45]

The overall degradation of land, which should be more accurately referred to as depletion of land, is a global problem with implications that affect agricultural productivity, the environment, food scarcity, and the quality of life.[46] Over the past fifty years, short-term economic gains by the meat and dairy industries have clouded the reality of poor and inefficient land management, failure to encompass sustainable practices, and long-term destruction of our topsoil. With more land being less productive, the narrow-visioned expansion of grazing and croplands into natural habitats occurs. As valuable topsoil is depleted by indiscriminate and continued use by livestock, attempts to restore this depleted land requires more and more of our natural resources, such as water, gas, oil, etc. Additionally, pollution is then invariably generated in the process, and there is increased use of chemicals, such as fertilizers, herbicides, pesticides, and even more of our energy resources. Desertification, another form of land depletion or degradation, also occurs as a result of over-grazing of livestock. This condition occurs when rich topsoil is lost to the point where minimal or no plant life can be supported. Desert-like conditions occur with lack of water retention, continued rapid erosion by wind or rain, and loss of nearly any form of productivity. Although these conditions are witnessed in many areas of the world due to the introduction of livestock, one of the more severe examples is in Africa. About 500 million Africans are affected by desertification, which seriously undermines any agricultural production, with livestock being the primary cause.[47] Modern livestock production and its effect on land use are driven by demand for livestock products. If there is less demand, there will be less livestock production, then less land used, and ultimately, less destruction, less topsoil loss, and less overall land depletion.

According to the Global Footprint Network (GFN), the global demand for land overtook global supply by the end of the 1980s. GFN was established in 2003 to measure human impact on earth. The organization comprises scientists, non-governmental organizations (NGOs), academics, and over ninety global partners that are universities and technological institutions, and it is supported by twenty-three countries. Their "footprint" is a data-driven tool of measurement that tells us if we are living within our ecologic budget, or if we are using natural resources faster than the planet can renew them. It's estimated that the human ecological footprint is currently 30 percent greater than the entire planet can sustain.[48] Moderate projections suggest that if consumption and depletion continue at current levels, we will need the equivalent of two earths to support us by the mid-2030s. Livestock and related processes are the largest contributors to our ecological footprint through land uses of grazing and feed crops, land needed to absorb CO_2, and even indirectly through fisheries needed to produce fishmeal for livestock feed.

The first level of land depletion is when forests are cleared for livestock. This reduces the soil's ability to absorb CO_2 and produce O_2 and also destroys water cycles. The second level of depletion or destruction is after natural ecosystems and land have been cleared; this is when the development of cattle ranching and pasture degradation related to overgrazing occurs. This pasture degradation results in soil erosion.

The livestock sector uses more than 9.6 billion acres, or 30 percent, of the world's surface land area. Certainly there must be concern for rapid global increase in the human population and the effect this has on our available resources. However, in terms of land use, it is our choice of food—collectively by our population and subsequent land management—that is having the most destructive effect. It is not simply a matter of having too many people for our planet; it is what these people are doing with our land and resources that is of most concern. One of many examples of this can be seen in our own country. There has been a population boom over the past thirty years in the southwestern United States; population there has doubled in this time period, with an increase of over 35 million people. Of the 99 million acres of land in this area of the United States that comprises the states of California, Nevada, and Arizona, 82 million acres are rangeland or pasture for cattle, while only 500,000 acres are classified as urban, being used as living areas by the population.[49] Although there are 16 million acres remaining as state, federal, or some private land, most of this will be taken over by livestock as our demand for meat continues to rise. The 82,000,000-to-500,000 ratio of land use is absurd and found during the time of a population spike. This example shows that our land is used primarily for cattle, not for humans. When combined with the destruction of land that occurs, this imbalance displays an obvious inefficiency of use.

The destruction, or global depletion, of natural habitats in order to establish grazing or cropland for livestock is, not surprisingly, the leading cause of biodiversity loss. Countless species of plants and animals are either extinct or severely threatened from our practice of raising livestock. Some 825 terrestrial ecoregions are identified by the Worldwide Fund for Nature across all biomes and all biogeographical areas on earth. Ecoregions cover large areas of land or water and are characterized by the natural communities of habitat and wildlife that they contain. In these regions are found characteristic and geographically distinct groups of various species of living things. The biodiversity of fauna, flora, and ecosystems that constitute and characterize an ecoregion also usually is distinctively different from that of another ecoregion. Each of these regions is classified by biome type, which is a distinct group of plant communities determined by climate and rainfall. For instance, various types of grasslands, forests, and deserts are all further distinguished by in which climate they are found (tropical, subtropical, temperate, boreal, etc.), which allows for specific grouping into biome type. Of these 825 major land ecoregions, 306 are reported to have livestock as a current threat. Another analysis by the World Conservation Union Red List of Threatened Species reports that most of the world's endangered or threatened species are suffering habitat loss due to livestock.[50] There is clear, unregulated, unnecessary continued proliferation of livestock into natural habitat to satisfy our demand for use of animals for food, without regard for the irreversible effect on biomes, ecoregions, and all the natural communities of living things they contain.

The most recent and comprehensive assessment of the effect of our activities on other living species took place in Nagoya, October 26, 2010, at the Convention on Biological Diversity. The meeting received minimal media coverage due to the minute-by-minute attention given to our financial crises. I found this both interesting and disconcerting, as the gathering in Nagoya was about the future of all life on Earth. The convention in Nagoya was a gathering of the world's leading authorities from sixty-two participating countries to report a summary of the current status of our planet's loss of biodiversity, as well as to create a plan to reduce the current rate of extinction. It was essentially a follow-up convention from the meeting held in 2002, where there was an agreement by the world's governments, as the Executive Summary and Report explains, "to achieve by the year 2010 a substantial reduction in the current rate of biodiversity loss ..." that has not been met. Of the five "principal pressures directly driving biodiversity loss—habitat change, overexploitation, pollution, invasive alien species, and climate change—all are increasing in intensity."

Species that were assessed for extinction risk are moving closer to extinction, with amphibians and coral reefs showing the most rapid decline. Almost 25 percent of all current plant species are nearing extinction. As of mid-2010, 170 countries had adopted national biodiversity strategies and plans of action. Delegates agreed to protect 17 percent of the land area of the world that remains and 10 percent of the oceans by 2020. Specifics of exactly how this is to be accomplished are vague at best, and these targets are not ambitious enough. A "flexible framework" was adopted that allows countries to continue doing as they wish, with no real threat of sanctions or repercussions serious enough to effect change. As of December 2010, the U.S. had not even signed the UN Accord on Biodiversity. Examples cited of existing scenarios that cause the most effect on biodiversity loss are: "Tropical forests which are being deforested in favor of crops and pastures, climate change and pollution, and overfishing which would continue to damage marine ecosystems and cause collapse of fishing populations." The two primary suggested solutions were:

1. much greater efficiency in the use of land, energy, fresh water, and materials to meet growing demand for food, and

2. use of market incentives and avoidance of perverse subsidies to minimize unsustainable resource use and wasteful consumption.

Because raising livestock is a primary cause of land depletion, global warming, water usage, deforestation, and pollution, doesn't it make sense that this topic should have been addressed specifically? And because eating fish is the cause of our exploitation of fishing stocks in our oceans, doesn't it make sense that this topic also should have been addressed? There should be no mistake that our choice of eating animal products as food is the driving force behind the demand to continue raising livestock and to fish. This, then, is obviously what fuels the continued loss of habitat and biodiversity on our planet. And yet nowhere in the Executive Summary of this very important gathering was there proper addressing or direct management for resolution. Let me help:

The major cause of biodiversity loss on our planet is from the livestock we raise for food and from overfishing of our oceans, both driven by demand. The most effective methods of correcting this are to:

1. Eat only plant-based foods.

2. Eliminate all subsidies for any businesses that produce animal products.

3. Create incentives for those businesses that produce plant-based foods.

4. Educate all nations as to the real reasons for biodiversity loss and the most effective and quickest way for resolution. Specifics are the rule; there is no room for generality.

There have been five previous massive extinctions on earth, with the last one occurring 65 million years ago at the end of the Mesozoic era, due in large part to the effect of the impact of an asteroid striking what is now the Yucatan Peninsula in Mexico. We are currently in our sixth period of extinction. The major difference between previous extinctions and the current one is quite clear—the current extinction is not due to external, uncontrollable forces; it is due to our own actions. We humans are causing the massive loss of species of plants, insects, and animals every day, and this has been happening at an escalated rate since the 1960s. Once more, it is not the number of humans as much as our combined actions with respect to how we affect Planet Earth. We are destroying our forests, especially our rainforests, converting natural habitats to pasture or crops for livestock, polluting and causing excessive global warming, using water at an unsustainable rate, and overexploiting many species of life, particularly in our oceans. Although there are many reasons for this, it is quite clear that the primary reason for the combined effect of human activities on loss of life is our choice of animals for food … period. Until clear communication and recognition is established regarding this reality, and then a concise plan created to implement strategies to resolve it, true progress will never be made. There *is no room* for a halfway approach. Heavy economic penalties should be imposed on all producers and consumers of animal products, as well as nations that continue the practice.

Food depletion is also ongoing and occurring at an alarming rate. What is food depletion? Well, let's look at the United States, for instance, where an almost unbelievable 70 percent of all grain produced is fed not to humans but to animals that are raised for food.[51] The World Hunger Organization reported that six million children died of starvation in 2009 alone. Another one billion people currently are suffering from hunger and malnutrition. There is more than enough grain produced each year to eradicate world hunger, but the solution is to *stop giving grain to livestock* and to simply give it to those who are starving to death.

In 1986, during the food crisis in Ethiopia, there was an increase in global awareness of hunger in that country—most media services covered the topic well, including using infomercials on television. What was blatantly lacking in media coverage, however, was the fact that each day, while thousands of people were dying from hunger, Ethiopia was—at the very same time—using a significant amount of its agricultural land to produce cereal grains (linseed, rapeseed, and cottonseed) for export to the UK and other European nations, to be used as feed for European livestock.[52]

Today, as then, millions of acres of undeveloped third-world land are being used exclusively to produce feed for European livestock—and those livestock eventually end up in the United States. I find it interesting and tragic that 80 percent of the world's starving children live in countries where food surpluses are fed to animals that are then killed and eaten by more well-off individuals in developed countries. It is estimated that one-fourth of all grain produced by third-world countries is now given to livestock. This figure has tripled since the 1950s.[53]

Each year in the United States, inefficient use of land for food is exemplified by the fact that 157 million metric tons of cereal, legumes, and vegetables—all suitable for human use—is fed to livestock to produce barely 28 million metric tons of animal protein for human consumption.[54] Globally in 2007, there was a "record harvest," with 2.1 billion tons of grain production.[55] There should not have been much difficulty, then, with providing assistance to those people in the world who suffered from hunger—except for the fact that over 50 percent of all crops grown were used to feed livestock instead. Each year nearly, one billion tons of grains and vegetables are fed to animals in the meat and dairy industries.[56] We have enough of the right type of land on this planet to provide healthy food for humans in a sustainable manner, but currently, the land is being depleted—perhaps irreversibly—by livestock operations and unsustainable agricultural techniques used to produce feed that supports animals to be slaughtered, instead of its going directly to humans to keep them alive and healthy.

CHAPTER VI

Water and Oceans

Part 1: Drinking water and sustainability—where is it all going?

Part 2: Our oceans—what is happening below the surface?

"When the well is dry, we know the worth of water."
—Benjamin Franklin

THERE ARE THREE PRINCIPLE AR-

eas of global depletion as it relates to water: the realities of our drinking water and who is using it; the current exploitation of our oceans and fish; and the pollution of both of these. Let us begin by taking a closer look at our drinking water.

Part 1: Drinking water/fresh water

Once again, we are raising 70 billion animals each year that are killed and eaten. Each of these animals, particularly cattle and pigs, require much more water daily than either you or I would—many, many more gallons. Now, multiply that times 70 billion. It is not so difficult to surmise where all of our drinking water is going. It actually requires over five thousand gallons of water to produce one pound of meat.[57] It is well known that each of us needs six to eight eight-ounce glasses of water each day for proper hydration and proper health. Were you aware that a cow needs thirty gallons of water each day, just to stay alive, and a pig needs up to twenty-one gallons?[58] That is *up to sixty times* more water for just one animal than we humans would drink in one day. And that does not count all the water it requires to wash the areas where the animals live or that is used in their slaughtering and butchering processes. Are you asking, "So what?" Well, why would anyone want to give sixty times the amount of water that we would drink ourselves in one day to pigs or cows or any other animal that we then kill a year later and eat parts of it—parts that are, factually, unhealthy for us?[59] Remember that much of the water that is used for these animals is to produce crops for livestock feed, drinking water, or slaughtering is, in most cases, nonrenewable in our lifetime.

To understand what can only be viewed as bizarre misuse of our drinking water, it is important to know that there are currently eight million people living in the states of Iowa and Missouri, amid more than 50 million pigs.[60] That's 50 million pigs, *in just two states*, that are raised each year, using our land, crops, fossil fuel, and water, with massive amounts of waste polluting our ground, water, and air. Again, so what? Well, that is where your bacon, pork chops, hot dogs, etc., are really coming from, not from your local grocer. Every time you eat bacon, pork chops, hot dogs, or any other animal product, please keep in mind all the depletion for which you, personally, are responsible. Don't look the other way or pass the blame; it is you—and you alone—who ate that particular animal, thus keeping the wheels of depletion moving onward. If you did not ask for and eat it, the meat and dairy industries would not produce it and exploit our resources. There are alternatives—many alternatives.

In fact, more than one-half of all the water used in the United States is, in one way or another, given to livestock. That means one-half of all the water used by all humans and businesses is going to animals raised for us to eat.[61] Consider that it requires more than five thousand gallons of water to produce one pound of edible beef, but only twenty to sixty gallons to produce one pound of vegetables, fruits, soybeans, or grains (all of which are more nutritious than meat and which do not contribute to a number of diseases, as do animal products).[62] Essentially, one person can save more water simply by not eating a pound of beef than he could by not showering for an entire year. It does not seem logical to continue in the direction we have been going with regard to raising animals for food.

It must be understood that fresh drinking water from the ground is not infinite in quantity, and it is *not renewable* in our lifetime. Freshwater resources are scarce—just 2.5 percent of all water on earth, and 70 percent of that is locked in glaciers, snow, and the atmosphere. This leaves accessible fresh water at less than 1 percent.[63] While some water is replenished through the natural evaporation/precipitation cycle, much is gathered from underground aquifers or surface water, such as rivers and streams. Depletion of our freshwater supply occurs globally, due to excessive withdrawals for agriculture, as well as poor water management. Whether using surface water or groundwater, agriculture accounts for 93 percent of water depletion worldwide, with the majority used for irrigated crops for livestock.[64]

We must also redefine which water resources are really "renewable" and which are clearly not. In the United States, there are common misconceptions:

- You do not have to worry about water; it's just there for you whenever you want.
- Use of our surface water is mostly by industries.
- Underground water supplies are renewable.

Actually, depletion at all sources is occurring at a rate that simply cannot be sustained.

The mighty Colorado River—1,400 miles long, with up to 24 million acre feet of annual flow—is one of the largest and longest in the world.[65] It has been such a force that, over time, it created the Grand Canyon on its way to the Sea of Cortez. Until 1936, when the Hoover Dam was constructed, this river continued its natural path to the ocean in Mexico, where it formed the two-million-acre Colorado River Delta. This delta was once one of the largest in the world, supporting a large population of plants, birds, fish, marine mammals, jaguars, and deer, as well as the descendants of Native Americans who had lived there for over one thousand years.[66] Today, this great delta does not exist; freshwater flows no longer reach it. Unbelievable as it might seem, the mighty Colorado River ends in the desert, some two to three miles from the sea, and has only reached the ocean sporadically over the past eighty years.[67]

This happened because Americans needed water to fuel

their activities in the very dry western states; most of these activities were related to raising cattle. In the process, the United States claimed—and now retains—95 percent of the river's water and has built nine major dams and eighty diversions on the Colorado River that control the water to the point where only a trickle makes it to the point where it once roared mightily out to sea.[68] The water that does eventually make its way to the Mexico desert is now heavily laden with chemicals, such as pesticides and fertilizers, which is runoff from all the alfalfa fields in California and Arizona—alfalfa fields that are producing feed for cattle operations. The Colorado River has had its water tapped, mostly for livestock, with the belief that it is renewable and will not run out.[69]

There are two common misconceptions regarding water use of the Colorado River and in the western United States in general. One is that water is renewable. To a point, it is not. The second misconception is that the rapidly growing urban areas (Las Vegas, Phoenix, and many others) are the primary users of this river's water supply. They are not. Most of the water has been and is still being used for livestock. In the Upper Colorado River states, such as Wyoming, Colorado, Utah, and New Mexico, 90 percent of the water use of the Colorado River is irrigation of crops, leaving only 10 percent for urban and other uses. Eighty-eight percent of this irrigated land is for crops fed to livestock.[70,71] Similar percentages are found in the lower Colorado River Basin states, such as California, Arizona, and Nevada, with nearly 80 percent of all river water used for livestock. So while enormous public cost has resulted in the area's massive water damming and diversion projects, the Colorado River Basin states and portions of the states just outside the basin are not receiving much of the river's water, while the river itself is slowly drying up.

Studying the expansive eight-state High Plains area of the United States also provides an example of how our demand for eating animal products has affected drinking-water use and just how ridiculously it has been managed. A portion of this area has been termed the "Dust Bowl," due to how dry it is and its susceptibility to repeated episodes of drought. But nearly half of the cattle in the United States are raised in the High Plains states and rely on one aquifer system. With average annual rainfall of less than twelve inches, it is not an area that could naturally support the growth of crops. Since the early 1960s, and with technological advancement related to pumping, farmers have irrigated the dry land, which dramatically increases crop productivity. Low-budget, small-scale local irrigation systems from an underground water supply just a few feet below the surface—an aquifer called the Ogallala—allow for more crop growth without regulation.

While different types of grain are produced in this area, the majority is corn. The difficulty is that 93 percent of the grain grown in this area and in America is used to feed cows, pigs, and chickens that are then killed for us to eat.[72] And that is not where the inefficiency ends, as water from this aquifer is also used in the slaughtering and processing of the animals. This scenario is also found elsewhere in the world, with freshwater supplies used primarily for livestock. Although direct annual water use by livestock worldwide is estimated at 23 percent, there is a recognized inability to properly quantify the real global depletion impact of the livestock sector. This figure could very well be much higher, because most water loss is unreported or is difficult to measure, as it is used and depleted throughout the livestock chain. For instance, massive amounts of unreported water are lost due to evapotranspiration by feed crops, use in chemical application, and various washing areas, transportation, processing, and packaging.[73]

In California, 42 percent of irrigation water is used to produce feed grain or drinking water for cattle and other livestock, which has caused such a drop in water tables that some areas of land in the San Joaquin Valley have sunk by as much as twenty-nine feet. There, aquifers are pumped at rate that exceeds recharge by more than 500 billion gallons annually.[74]

At Iowa Beef Processors, just one of many slaughterhouses in that state, over four hundred gallons of water are used to kill and process one cow.[75] That plant alone slaughters over five thousand cattle per day, or 1.5 million each year. This requires that one Iowa plant, which processes a fraction of the world's livestock, to use 600 million gallons of water, which is pumped from the Ogallala aquifer in a single year.[76] This is nonrenewable, pristine drinking water, formed from glaciers thousands of years ago. The Ogallala, which is one of the largest aquifers on earth, supplies water to eight states and is a striking example of the unnecessary depletion caused by the demand for animal products for food. By 1990, it was drawn down by three to ten feet per year, a rate many feel will deplete it entirely by the year 2020.[77] Like many underground water supplies worldwide, and with recharge rates of less than a half-inch per year, the water of the Ogallala, which was formed in the ice ages, surely will be gone soon—unless there is drastic reduction in our dependence on it for raising livestock. Once it is gone, it is gone.

The solution to the declining freshwater supply does not lie in increasing demand and then simply supplying more, as we are currently doing. Rather, it is to understand where our water is going, to reduce this extraneous demand, and then to change to a more efficient use system. Knowing that the majority of our water supply is involved in the raising and processing animals for food, and knowing that it requires a mere fraction of water to provide a more nutritious food derived from plants, one of the solutions to global depletion of water is quite simple: *Stop all consumption of animals for food*. Pretty simple.

Part 2: Oceans

So let's go back to that restaurant where you ordered a burger, steak, or another form of animal product. And let's presume, in this new scenario, that you now feel enlightened in some capacity to think that some animal products eaten as food are less healthy for you than other animal products. Specifically, you have decided to cut back on red meat—which is an interesting term, in that this form of animal tissue has a generous blood supply; hence, it appears red in color. Back at the restaurant, what do you order? Fish, of course, because you have heard that this form of animal product is healthier for you, with its good type of fat. Actually, fish of any type has both saturated fat and cholesterol, both of which are not necessarily that healthy for you to consume. An ongoing Environmental Protection Agency (EPA) study found that an alarming 100 percent of freshwater fish samples from the United States contain mercury, and a large percentage of certain fish caught from the ocean contain heavy metals and/or polychlorinated biphenyls (PCBs), which are highly toxic, cancer-causing chemicals.[78] Additionally, not one fish caught anywhere in the world has fiber, antioxidants, or phytonutrients, all of which make you a healthier person; these can only be found in plant foods. While there has been much hype about omega-3 fatty acids, no hype has been given to plants that can provide a more stable form of omega-3 fatty acids than is found in fish (and is broken down easier by your body). Plants such as flax or hemp contain omega-3 fatty acids in a quality and quantity whereby it is easy to obtain the needed daily amount. And unlike the baggage that comes with fish, these plants come to you with much needed fiber, no cholesterol, no saturated fat—and no loss to our oceans. Let's examine one more very important issue: where did your fish come from? You may say, "It doesn't matter"—you are simply eating lunch, so who cares? Unfortunately, the fish you are eating and the demand for fish by many other people have placed such a burden on our oceans that they may never recover.

Truthfully, our oceans are a mess. The rate of depletion, human-induced extinction, and environmental degradation in our oceans is most likely "greater than anything witnessed on land," as described in a recent report by the United Nations Environment Programme.[79] Coastal zones, as well as the high seas, are under great stress from unsustainable practices, including overharvesting of fishing grounds, bottom trawling, pollution and dead zones, and infestation of invasive species—all fueled by the massive desire to eat more fish. While global warming trends are measurably affecting our oceans, it is the indiscriminate, unregulated overfishing of our seas that will have the most profound and long-lasting effect on all of the complicated intertwining of fragile ecosystems. Our oceans are highly complex and dynamic systems, all interconnected to each other and vital to all living things on earth. The core of these vital systems and environmental mechanisms is living marine biodiversity itself.

The largest amount of marine life is associated with the seabed, especially on continental slopes and shelves of seamounts. Seamounts (mountains rising from the ocean floor that do not reach to the water's surface) are home to corals, sponge beds, and numerous communities of species. They provide feeding, spawning, nursery grounds, and shelter for thousands of species of commercial fish, as well as migratory species such as whales.[80] Many seamounts that are separated from each other behave like marine oases, with distinct species and communities found nowhere else on earth, such as the Tasman seamounts, with a 34 percent *endemism rate* (sea-life species that are found nowhere else on earth). With traditional fishing grounds now depleted, the fishing industry is targeting newer stocks, with more sophisticated locating equipment, farther offshore, including around and on seamounts. Large industrial vessels and fleets operate for weeks and months, targeting deep-water species on continental slopes and seamounts. Over 95 percent of the damage—and possible irreversible change—to seamount ecosystems is caused by unregulated and unreported bottom-fishing, with extremely destructive gear such as trawls, dredges, and traps. It is estimated that trawling alone is more damaging to seabed areas than all other fishing gear combined and is destroying deep-sea communities that will take decades and centuries to recover—if at all. These species and ecosystems are particularly at risk with additional stress, such as climate change and pollution.[81]

What happens, essentially, is that fishing vessels clear a seamount area of as much fish as possible, and once devastated and depleted, fishermen simply move on to the next seamount to start the process all over again. Many known seamounts are already overexploited to the point where extinction may well soon follow or recovery may take centuries.[82] "Extinction" and "recovery" are terms that, in this context, need to be applied to a particular marine species that is dependent on seamounts, such as the right whale, and to various complex ecosystems—those with which exploited marine species are involved. We do not even fully understand many of these species and marine ecosystems, much less appreciate them.

There are four levels of what I consider classic human behavior with this demand for fish and subsequent overfishing situation:

1. The human population consumes a food item, but they do not have a clue to where it comes from, what it took to get it to their mouths, or what it is doing to our planet; or they do know but simply do not care. Either way, this behavior is irresponsible as well as unacceptable.

2. The fishing industry is oblivious or does not care and is blinded by the immediate, short-term economic gain, which is also not acceptable.

3. Those who are somewhat aware are not addressing the issue aggressively enough and are not solving the problem quickly enough. These are organizations and individuals, such as UN Special Report committee members and others who have studied this topic, as well as those in a position to make policy change. Again, this is not acceptable, as depletion continues to occur on a daily basis.

4. We are demanding, consuming, and ultimately irreversibly destroying living species and interconnected ecosystems, *the complexity of which we, as humans, do not even fully understand.* It is the essence of ignorance, the "me kill you" mentality that should have been left back in the Early Pleistocene era.

These fishing areas of exploitation are beyond national jurisdiction and have limited, if any, regulation, which makes them extremely vulnerable to further exploitation beyond recovery. Most deep-sea fish in these areas are slow to mature, produce only a few offspring in their lifetime and, therefore, will more easily be completely destroyed by heavy bottom-trawling and other gear now used.[83] Of the seventeen primary fishing stocks worldwide, *all* are either overexploited or on the verge of collapse. Examples of commercially extinct areas are the Grand Banks near Newfoundland and the Georges Banks off New England, both once considered the most productive on earth. Until fifteen years ago, the fish you ordered for dinner would have most likely originated from one of these two very productive areas. At less than 1 percent their original numbers in these waters, now there simply are no fish.[84]

Across all our oceans, the Food and Agriculture Organization of the United Nations (FAO) estimates that 70 percent of the world's fish species are either fully exploited or depleted.[85] The World Conservation Union lists 1,081 types of fish worldwide as threatened or endangered.[86] There are currently four million fishing vessels that catch fish at a rate and amount that is almost three times that considered to be sustainable.[87]

Our most competent ocean scientists admit to understanding and comprehending only a very small amount about our seas, so it is interesting that someone had enough unilateral confidence in himself to come up with the figure of what is "sustainable" in any ecosystem in our oceans.

Although latest statistics reveal that a record 106 million tons of fish were caught in 2009, the report does not account for the millions of tons of total sea life caught in the process and discarded, either dead or dying.[88] This "bykill" is most pronounced with shrimp fishing. Please think about this just for a moment when you next eat shrimp: for every one pound of shrimp sold and consumed, more than twenty pounds of other sea creatures are caught and killed in the process. Innocent victims include fish that has no commercial value, juvenile fish, turtles, seabirds, and marine mammals, such as the dolphin.[89]

Overfishing and the massive amounts of subsequent bykill have such detrimental global implications that the United Nations has adopted resolutions to help in the reduction of both. The real difficulty, however, lies in the adoption and actual enforcement of policies by the international community. For example, one such UN ruling (Item 150, Report of the Secretary-General, July 14, 2006) states: "Observers have collected data on fish discarded at sea by most vessels. Fishermen are required to properly release, to the extent practical, unharmed sharks, billfishes, rays, Dorado, and other non-targeted species, including sea turtles, and to receive some training in release methods."[90]

So with this ruling, are we to understand that there will be "*observers*" of "most" vessels? That fisherman are required and "*reminded*" (but by whom?) that they are to "properly" release, "to the extent practical," any "unharmed" species, and that they will receive "some training in release methods"? Might we assume, then, that the problem is now solved? Of course not. The policy itself is ludicrous, and there is *no assurance whatsoever* by our governing bodies that any part of the problem will be solved. While I am encouraged by the recognition that there is a significant overfishing and bykill difficulty, proper implementation and enforcement of appropriate laws quite clearly remains problematic, at best.

It is also very important to note that one-third of all fish caught worldwide are used as fishmeal—to feed the ever-growing numbers of livestock.[91] That is correct: your desire to eat animal products (fish *or* livestock) fuels industries that are depleting our planet and destroying our oceans and environment. As executive director Achim Steiner stated in a recent report of the United Nations Environment Programme with regard to our oceans: "We are now observing ... in the absence of policy change. a collapsing ecosystem [of our oceans] ... with climate being the final coup d'grace."

Depletion of our oceans is very real and, in many cases. is occurring at an irreversible rate. This catastrophic depletion is initiated and perpetuated by our demand for fish, the "healthy" alternative—but healthy for whom?

CHAPTER VII

Pollution

Yes, that would be you

"The activist is not the person who says the river is dirty.
The activist is the person who cleans up the river."
—Ross Perot

EVERY SECOND OF EVERY DAY, THERE

is depletion in the form of pollution. Pollution of any kind depletes the environment of clean healthy soil, waterways, groundwater, and the air we breathe. Some of the largest contributing sectors to this pollution are the meat, dairy, and fishing industries—and those who choose to eat things that these industries produce. "How can that be?" you say. "I simply eat it; I am not polluting." Well, yes, you are. And here is how it all works: Your contribution to pollution begins with what you decide to purchase to consume. It's not just with the occasional purchase; it's with every food item you eat, every day. With meat and animal products, the pollution associated with your choice is massive. In order to raise that animal for you to eat, there is baggage that silently comes along with it—silent to you, that is, although it speaks loudly elsewhere. In the United States alone, chickens, turkeys, pigs, and cows in factory farms produce over five million pounds of excrement *per minute*. These are the animals raised each year so that people can continue eating meat, and they produce 130 times more excrement than the entire human population in our country. This manure sewage is responsible for global warming, water and soil pollution, air pollution, and use of our resources. The waste produced by the animals raised for food includes with it all the antibiotics, pesticides, herbicides, hormones, and other chemicals used during the raising and growing process. Accompanying this is methane released by the animals themselves, as well as the carbon, nitrous oxide, and additional methane emissions produced during the whole raising, feeding, and killing process.

Regarding pollution of our global water supply, livestock are responsible for 37 percent of pesticide use, 55 percent of erosion, and 50 percent of the volume of antibiotics consumed.[92] This ultimately ends up in our waterways, either directly or through runoff, creating water contamination. Livestock are responsible for 33 percent of the nitrogen and phosphorus loads found in freshwater resources.[93] While there is no current assessment of the effective load into freshwater resources of sediments of heavy metals or biological contaminants, it can be reasonably assumed that livestock have a major role in these processes of pollution as well. In the United States, recent EPA studies have shown that 35,000 miles of rivers in twenty-two states and groundwater in seventeen states has been permanently contaminated by industrial farm waste.[94] Raising animals for us to eat pollutes our waterways more than all other industries combined.[95]

Pollution from animal factories is destroying our oceans as well. Streams and rivers carry vast amounts of excrement and chemical waste from livestock farms, which finds its way to the ocean. Deposits of animal feces, fertilizers, and toxic waste cause death of plants and sea life, as it causes massive algae populations that leave inadequate oxygen for other forms of life. One of the world's largest "dead zones" can be found in the Gulf of Mexico off the U.S. coast. This is an area about half the size of the state of Maryland, in which nearly all the sea animals and plants have died. A 2006 report by Princeton University concluded that a shift away from meat production would dramatically reduce the amount of nitrogen carried into the Gulf, to levels that would render this dead zone "non-existent."[96] The UN now reports there are 150 dead zones in the world's oceans, caused by an excess of nitrogen from farm fertilizers and sewage. Another example is the serious loss of marine life and ecosystems in the South China Sea due to nitrogen and phosphorus runoff, where it is now known that livestock are the primary inland source of contamination.[97]

Knowing that many fishing areas are becoming devoid of fish, the development of fish farms has exploded in the past few years. Because stocks of most of the top ten fish species are depleted and overexploited, businesses and governments have turned to other ways to produce fish. Aquaculture, the growing of fish in a farmed area, increased by more than three million tons from 2006 to 2007 and is expected to grow faster than all other animal food sectors.[98] This growth in aquaculture is driving an increase in global fishing, due to the need for fishmeal and fish oil, which is used in fish farms. It is a bizarre, ecologically unhealthy circle, where the demand to eat fish has taxed the oceans so there has been a proliferation of controlled fish-farm production, which places further stress on the oceans because of the need for fish-meal and oil in the production process.

These fish farms now greatly contribute to water pollution on two levels. The first is by further concentrating toxin levels and creating a higher potential for our exposure to them. When fishmeal and fish oil are used in aquaculture, the process concentrates carcinogens such as dioxins. This occurs because various contaminants and chemicals are found in many types of fish, which are then passed on, in more condensed forms, as they work up the food chain. Farmed salmon, for instance, consistently have much higher levels of dioxin than their wild counterparts. This is because they are fed a constant diet of fishmeal, which now has concentrated amounts of the many pollutants to which all the fish comprising the meal were exposed during their lifetime. Farm-raised salmon and other fish now dominate certain markets, such as on the West Coast of the United States.[99]

The second level of pollution for which fish farms are responsible is the massive amount of waste they produce, which further depletes our oceans. Farmers confine thousands of fish into tiny enclosures in the ocean, with enormous amounts of feces and other waste being created and deposited into our waterways. Farms typically grow up to 90,000 fish in a pen that is 100 feet by 100 feet. In one area with adjoining pens, as many as one million fish can be grown at one time.[100] Organic and chemical wastes from these farms occur from all of the following sources:

- Fish feces and bodily waste products
- Fish mortalities that sink to the seabed
- Fish blood from farms that kill and bleed fish on site prior to sale
- Uneaten food
- Eight types of antibiotics
- Feed additives and coloring agents to make their white flesh appear pink
- Zinc and copper
- Paints and disinfectants

All of these produce increased nitrogen levels and toxins that cover seafloors beneath these farms, creating large dead zones.[101] One study in 2001 revealed that the farmed salmon just in British Columbia for one year produced as much nitrogen as the untreated annual sewage from 682,000 people.[102] A professor of fisheries at the University of British Columbia has noted: "They [fish farms] are like floating pig farms."[103] Antibiotic use is common, as well as pesticides and copper sulfate, an algaecide. This is because diseases and parasites run rampant in the cramped, overcrowded fish-farm conditions. Swarms of sea lice actually end up being inadvertently incubated on the captive farmed fish and then attach themselves on wild salmon and other fish as they swim past the farms in the ocean.

While fish-farm pollution has been studied and documented in North America, accurate reports have not been available for other areas of the world, such as China, where the industry has exploded. China produces over 70 percent of the global supply of farmed fish.[104] It can be safely assumed that along with that is the production of massive amounts of ocean depletion of marine life and pollution. Fish farms exist because we have depleted our oceans of numerous species of marine life.

The solution to loss of marine life is to eliminate the demand for fish, not to create new aquaculture industries that create an even greater level of depletion from additional fishing and pollution.

Whether discussing land, water, or the air we breathe, our food choices heavily affect the level of pollution and ecological sustainability. The more you choose to eat animal products, the more you contribute to the worldwide pollution. It really does not matter whether it is in the form of livestock or fish; there still will be an excessive and unnecessary amount of depletion and pollution.

CHAPTER VIII

Why Do We Do It?

A word about nutrition—do you really care?

"We are not just killing Mother Earth; we are killing ourselves.
Earth will be here long after we are gone."
—Blackfoot tribesman

CERTAINLY, THERE IS NO MYSTERY

as to whether meat, dairy, and food derived from animal parts is
good for your health; it is not. This is not just my opinion. Meat is
factually not healthy for you, and there is an exhaustive amount
of peer-reviewed literature that supports this. Additionally, numerous health organizations, such as the American Dietetic Association, the American Cancer Society, the American Heart Association, Physicians for Responsible Medicine, and many others,
all recognize the health benefits and advantages of a plant-based

diet and have supporting statements to indicate this. Then the question remains, why are the vast majority of people still eating meat? Why do we do it? The answer lies in a complex web of interactions that results in a continued tunneled belief that it is still healthy for you, despite the facts. You may wonder why I've included a chapter on nutrition in this book, which is meant to be about global depletion. The answer is simple: no book on global depletion would be complete without some mention of how the food we choose to eat causes profound depletion of our health.

First on the list is heart disease—the number-one cause of death in the United States—which accounts for more than one million heart attacks and 500,000 deaths every year.[105] Many studies have found that lifelong vegans have a nearly 60 percent reduced risk of death from heart disease.[106] The American Dietetic Association has declared that a vegetarian diet reduces the risk of many chronic diseases and conditions, not only heart disease but also including cancer, obesity, hypertension, and diabetes.[107] They further conclude: "Vegetarian diets offer a number of nutritional benefits, including lower levels of saturated fat, and cholesterol, as well as higher levels of fiber, magnesium, potassium, folic acid, and antioxidants, such as vitamins C and E and phytochemicals. Vegetarians have lower death rates from heart disease, lower blood cholesterol levels, lower blood pressure, and lower rates of hypertension and type 2 diabetes, and lower prostate and colon cancer." It is now widely known that vegetarian diets can even *reverse* heart damage already present.[108]

Then, there is cancer. The American Cancer Society (ACS) has as its number-one recommendation, on nutrition for cancer prevention, to eat a diet "with emphasis on plant sources."[109] This is supported by numerous studies that show that individuals who do not eat animal products have a 50 percent less likelihood of developing many cancers. The ACS and researchers at Yale have found that meat-based diets can cause cancers of the stomach, esophagus, colon, and prostate, and lymphoma.[110] People who eat hot dogs, sausages, and other cured meats have a 70 percent increase in pancreatic cancer.[111] The World Cancer Research Fund goes further and recommends a plant-based diet, listing fruits and vegetables as "convincing/probable risk reducers for cancer of the bladder, breast, cervix, colon, endometrium, esophagus, kidney, larynx, liver, lung, mouth, pharynx, ovary, pancreas, rectum, stomach, and thyroid."[112] Clearly, animal products eaten as food significantly increase one's risk of numerous disease states, such as the three largest causes of death—heart disease, cardiovascular disease, and diabetes—as well as many forms of cancers.[113] Eating meat and dairy also increase one's chances of contracting kidney stones and kidney disease, gallstones, and osteoporosis.[114] The obvious effect of all of this concurring information is the need for a change in dietary choices.

While there is much to be said about all nutrition issues as they relate to food choices and risk factors, let's look more closely at the relationship of the consumption of meat and dairy products to osteoporosis, as this represents an area of gross public misinformation and a subsequent state of being unaware. Despite what the dairy industry wants you to believe by its massive multimillion dollar ad campaigns, milk and milk products *do not* "build strong bones," and they will not prevent osteoporosis. In fact, numerous studies have shown that it is more a problem of limiting calcium loss than it is of increasing calcium intake. Countries that consume the highest amount of dairy products, such as Switzerland and the United States, have some of the highest incidence of osteoporosis, while other countries, such as in Africa and Asia, where virtually no dairy products are consumed, have the lowest rate of osteoporosis and hip fractures. While genetics and hormonal interaction may have roles, the principal reason for these findings may reside in the simple truth that individuals who consume high amounts of animal protein and dairy products are at a risk of depleting their calcium stores, regardless of how much calcium they consume. Animal protein found in all meat products has sulfur-containing amino acids, such as methionine, and large quantities of phosphorus, both which have been found to impair calcium balance. While you may think eating more protein in the form of meat and dairy is healthy, it actually is not. Your body cannot store any excess protein, and it must be excreted by the kidneys. During the elimination process of this, and in the attempt to balance excessive phosphorus from animal sources, calcium from your body is needed and used; thus, the unlikely loss begins. Beef and chicken, for instance, have phosphorous-to-calcium ratios of 15:1, while most vegetables have ratios averaging near 1:1, allowing for a much healthier calcium balance. Green leafy vegetables, such as kale, have quite a bit of calcium, which is as absorbable as that of meat and dairy products but without the baggage of excessive phosphorus and sulfur-containing amino acids. Eating kale may not provide you with a "milk mustache," and you may never be aware of its benefits through television or magazine advertisements, but it will create a healthier calcium balance for you and be much better for our planet than eating any meat or dairy product.

If animal products are killing us, why do we still eat them? The answer is quite frustrating and is wrapped in multidimensional levels. First, although there is a massive amount of supportive information in journals as ample validation, much of this information is passed over or overtly suppressed. Why? Because this information is controversial. In fact, I feel that this information is most controversial and damaging to the powerful industries and corporations that currently have the ability to suppress it—or to make life miserable for anyone who attempts to publicize it. Second, most individuals do not believe (or do not want to believe) that those meatballs that their mother or grandmother used to make were actually unhealthy or that eating them contributes to a number of debilitating diseases. To be honest, those very same meatballs (and similar animal products) consumed over a number of years may well have been one of the largest reasons for Grandma's death.

Milk—as much as you may want to believe otherwise—is not healthy for you either. A vast amount of evidence reveals problems with milk's protein, sugar, fat, contaminants, and lack of nutrients. Milk should no longer be recommended or considered required for growth or health benefits, as many organizations now recognize that it is unhealthy for consumption, due to the many health risks. It has been shown that milk contains numerous allergens, bovine growth hormones, and chemicals (herbicides, pesticides, fungicides, DDT, and others).[115] Some studies have documented as many as seventy-three contaminants found in any one milk sample.[116] It is also now understood that cow's milk causes asthma, food allergies, and chronic constipation, particularly in children.[117, 118] For these reasons, the American Academy of Pediatrics now recommends that no cow's milk be given to infants under one year of age.[119]

It is interesting to note that the majority of individuals in the world are lactose intolerant, meaning they are unable to digest milk and other dairy products. In 2000, findings in the *Journal of the American Dietetic Association* revealed that "approximately 75 percent of the world's population has lost the ability to completely digest lactose after infancy." For these individuals, consumption of any dairy products causes stomach upset, bloating, and distress. In the United States alone, those who are lactose intolerant include 83 percent of African Americans, nearly 90 percent of Asians, 60 percent of Native Americans, and 75 percent of Hispanics.[120] For these populations, milk should not be a dietary option, strictly from a physiological standpoint, as they lack the enzyme to digest it.

The answer to why we keep drinking milk lies in the fact that it has been, and still is, heavily ingrained in us by the dairy industry that it is a "health food" and is necessary for proper growth and bone development, which it is not. The original Food Guidelines and Pyramid that Americans use as a guide to proper nutrition was established by the Dairy Council and USDA years ago, with their own economic motives in mind. This very misleading guide was pushed by these organizations into every school system and home across America. Misleading marketing by the dairy industry still pervades today, with the "got milk" mustache campaign, and sayings such as "Milk gives you strong bones," and now, "You can even lose weight by drinking milk." As pointed out earlier, perhaps the public could be enlightened to the fact that milk does not give you "strong bones," as well as to all the documented ill effects that drinking milk presents.

The meat industry is even more involved with misleading the public that their animal products provide health benefits. This is despite the fact that there now is an enormous amount of medical and epidemiological studies that implicate animal products as a cause of cancer of the colon, rectum, stomach, prostrate, and breast.[121] Also, all meat has cholesterol and saturated fat, too much of which your body does not want or need. Many meat products, when cooked, have cancer-causing heterocyclic amines and polycyclic aromatic hydrocarbons.[122] Additionally, no meat of any kind will give you fiber, antioxidants, phytonutrients, or many vitamins or minerals that you need for optimal health. The scientific evidence against eating meat is indicting. Yet people still eat it because of cultural/social implications and misleading marketing. They simply cannot get themselves past the following hurdles:

- Not being given the correct information
- Being exposed to repeated misinformed messages in the media and advertisements, as well as having their mentoring physicians misguide them
- Having an awareness that all their friends and neighbors consume animal products
- Being emotionally constricted by the historical/cultural influence that their mothers, grandmothers, etc., ate this way

Cumulatively, these hurdles just become too much for the average person to overcome; it's simply too heavy to push aside and to do the right thing.

I believe most people truly do care about their own health, but because of lack of proper information, only a few of them also truly care about the health of our planet. Of these, still fewer people have open minds to the extent that they not only care about their health and that of our planet, but they also have the ability to be enlightened when encountering new truths. Now, of this very small percentage of the human population, only a few are willing to seek proper change. Along the way, this process is constrained by numerous social, cultural, and political hurdles. One of the primary reasons I decided to write this book was to provide truths that will empower more people by giving them the proper information, so that there will be an increase in numbers overcoming these barriers. Then, ultimately, change will take place.

Let's look briefly at the first step in this journey, which is caring about your own health. Many studies show that, in general, people care about their own and their family's health. A recent study by the Centers for Disease Control (CDC), however, found that people generally do not care enough about their health to actually implement change for themselves. The CDC 2009 obesity study, in particular, is alarming in three ways: First, the findings are in direct contrast to the public opinion that people do care about their health. Second, the statistics themselves indicate that Americans are becoming more obese and disease-prone. Third, and even more disconcerting, the report indicates that diet is the major contributor to health decline, but it does not elaborate on meat and dairy or their roles as principal factors to this relationship.[123] Obesity rates in adults have doubled over the past twenty years, while the rates for those between ages six and nineteen years old have tripled. Another shocking finding is that 25 percent of children, ages five to ten years, have high cholesterol, high blood pressure, or other early warning signs for heart disease, with one in ten teenagers having advanced fibrous plaques in their arteries.[124] The report further indicates that 70 percent of diseases and four out of six of the leading causes of

death in America are "diet-related."[125] This and other reports by
the Physicians Committee for Responsible Medicine (PCRM) and
the USDA conclude that eating less meat and dairy could prevent
yearly medical costs, ranging between $87 billion and $143 billion.[126]

What follows now is a simple overview of my journey and
experiences with finding healthy alternatives to fast food and the
many years spent researching the depletion of our health. At the
very onset of this experience in 1976, I began to understand the
reality of global depletion as it relates to food production and our
everyday choices.

From 1976 through 2000, I exhaustively researched the
fast-food industry, the eating patterns of Americans, the corporate objectives of the largest players (McDonald's, Burger King,
and Wendy's), nutrition, the ecology of our diets, marketing patterns, and the evolution of fast-food availability. I did this because I saw the desperate need for a new approach. It was obvious
that public demand for food in a quick-service venue would only
escalate as society became more and more fast-paced. It was also
quite clear that the current fast-food corporations that generated
billions of dollars in revenue could not care less about our health
or the health of our planet, regardless of their marketing fa-
çades. It became obvious to me that a healthy fast-food alternative business should be created—one with at least as many fran-
chised outlets as those already in existence from the "Big Three."
Right? If we, as a nation, can support 51,000 units that serve
unhealthy food, we should certainly and eventually be able to
support at least the same number of units serving healthy food.
The path seemed simple—an outlet should be created where only
the healthiest food possible would be offered and where education

and improvement of the health of our planet would be at the core of the business ethic. This was obvious. Well, disappointedly, I found out that although it was obvious to me, it certainly was not obvious to too many others.

From 1983 to 1998, I traveled around the country, speaking to CEOs, entertainers, politicians, and venture capitalists, explaining why this concept needed to be developed. During one presentation with a venture capital company, a member said, "Dr. Oppenlander, because you do not have an operational business to demonstrate that this would work, why don't you put up one yourself, and then come back to us for mezzanine [second level] money?" In 2000, I launched "Ope's—fast food the world can live with." This was a wonderful quick-service restaurant that offered just what was needed. It offered not only those foods that tasted delicious and were created in an artisan fashion using sustainable methods, but also that would be healthy for the customer and the environment. Through our restaurant, I wanted no saturated fat, no cholesterol, no hormones, no pesticides or herbicides, no heterocyclic amines, no inefficiencies or burdens on our environment, so no animal products were used—only vegan and organic. We had oil-less French "fries"; chocolate, vanilla, and fruit shakes; seven different burgers; six single-serve pizzas; and nine varieties of our trademarked "Stuffed Sandwiches." We also served salads, soups, and trademarked cookies. Everything was served to the customer less than five minutes after placing the order, and it was organic and sustainable. We developed team management and production protocols, and streamlined all food products and systems of operations. Additionally, proceeds were placed back into the business, with a large percentage donated to causes that would improve our environment globally.

We were privileged to have developed a loyal following of appreciative patrons, but my concept of having this business available for the entire world fell short—we simply did not have the number of people in our area at that time who appreciated this type of food. I learned that it was a matter of geography but also of enlightenment. Not enough people were at a level of understanding to really care about improving their health and the health of our environment. Initially, this was disappointing, but it provided the impetus for me to develop a deeper understanding of those mechanisms that affected public food-choice awareness and for me to help facilitate much needed change.

In 2002, I closed the restaurant to move attention to our production facility, which concentrated on producing our trademarked organic signature items: Ope's Organic Burgers, Ope's Organic Stuffed Sandwiches, and Ope's Organic Cookies. These items are sold to special retail outlets, hospitals, and universities. Along with their delicious gourmet taste, all of our Ope's items provide an opportunity to improve your own and our planet's health.

Through the combination of these experiences over the years, I have observed the interesting and very frustrating behavior patterns of Americans with regard to food choices, pathways of information and marketing of food, and the future perspectives.

Since 1987, I have lectured to numerous hospitals and school systems regarding the health benefits of a plant-based diet. Developing enlightenment with regard to food choices and creating change in those locations, although improving, continues to be challenging. Students, for example, are quite receptive to new information and how it might apply to them or our planet. Those in a supervisory position, however—those who actually can make proper decisions to invoke change—simply are not as receptive. They are either set in their ways and unwilling to become enlightened, or they are too passive and unable to commit, or they are overwhelmed by the political, social, and cultural issues such change may create.

All too frequently, I would conclude a presentation at a university, and a committee of ten or twelve student representatives unanimously would agree to incorporate the products my company offered. They recognized and appreciated that these products are plant-based and healthier for them and the environment. But weeks, months, or sometimes years later, none of our products would have been ordered, because the single individual who acted as food purchaser for the university would not take the steps to change. This was primarily because the purchaser was out of touch with the benefits of this type of food or was politically influenced by larger food providers. Sadly, this scenario was found repeatedly at most universities in Michigan and elsewhere across the Midwest. Interestingly, these are exactly the locations where healthier food products and information should be provided. Why? Because these same students are our future leaders and change-makers, and because the Midwest—and Michigan, in particular—is the area of the country where you'll find the some of the unhealthiest states, with alarming rates of obesity, adult-onset diabetes, heart and cardiovascular disease, and some diet-related forms of cancer.

Alarmingly, the situation is the same with hospitals, where you would think only the best diet, with the most up-to-date science behind it, would be available. Given the revered status in which we have placed these institutions and physicians, certainly they should be doing the right thing with regard to diet and your health—but they are not. Some of the unhealthiest food available is offered and supported by physicians and dieticians and is found in all hospitals. Additionally, some of the most unaware and narrow-minded indi iduals in decision-making positions regarding food choices are found within the hospital setting. The frustration was never greater than when I found myself repeatedly offering our food products at the University of Michigan Hospital, products that were requested by the vast majority of their medical students. But the administrators and food purchasers simply could not grasp the idea or move forward with a food item that would be the healthiest offering for their customers and the healthiest for the planet, and which was their first 100 percent organic product. They also could not imagine how these new items could be procured outside of the normal chain of business vendors.

Ultimately, most hospitals did briefly offer our products but only after *years* of my meeting with them, and after years of many vocal and enlightened medical students insisting that change was in order. The irony of this situation is that while I was struggling with educating and convincing the administrators that ours (or similar organic, plant-based products) were necessary, the hospital at University of Michigan continued to promote and support an in-building unit of Wendy's! You read that correctly. The University of Michigan built and promoted a Wendy's franchise in their hospital, adjacent to their main cafeteria, for all their students, faculty, patients, and visitors, while at the same time struggling to justify purchasing the organic and healthy food items that my company could provide.

Today, the University of Michigan has evolved to the point where the Wendy's unit has been eliminated, but there continues to be a severe inadequacy in providing truly healthy foods for its students, staff, faculty, and visitors. This inadequacy is fueled by a vivid dysfunction in the systems involved with their food procurement. In early April 2010, I met the Director of Food Purchasing for the entire university to discuss the disparity in what the students needed and wanted with regard to food choices and what they really were receiving from the university; I also provided a proper base of enlightenment for them. As I entered the hallway in a building on campus that led to the room where we were to meet, I stepped over a three-foot diameter rubber poster embedded in the floor, which stated: "Be part of the Blue Planet Movement"—this was a campus-wide initiative that encouraged the UM community to do things daily to improve the environment. At the same time, on the homepage of UM's website, there was a photo of their university president, Mary Sue Coleman, with the following message:

> The University of Michigan takes its responsibility of protecting and preserving resources very seriously, and every contribution can make a difference. I challenge everyone in our community to think about how even the smallest efforts will work to make our great institution even greener.

> —*U of M president, Mary Sue Coleman*

> **Save energy. Save the planet. The difference starts with you!**

Great message, right? Well, let's look more closely at this. Following a more than hour-long discussion with the director, this is what I learned:

- He did not understand that food choices play one of the largest roles in the global depletion of our resources.
- He did not know that eating plant-based foods requires substantially fewer resources than eating animals or that plant-based foods are generally much healthier for the students. (In fact, he was eating a burger from McDonald's as we began the meeting.)
- There is no program in place to provide ongoing education for the Director of Food Purchasing, his staff, the educators, or administrators of UM regarding the role of food choice in nutrition and sustainability.
- He thought the word "sustainability" meant "nutrition."
- He did not understand the ecologic or general health benefits of foods grown organically, and there is no program established whereby organic foods are even considered for purchase.
- He was confused as to the concept of buying locally. To my question of "Do you feel there are reasons and benefits for you to purchase food produced locally or by Michigan businesses? And therefore do you and the UM have any specific programs in place to accomplish this?" The response was, "Yes, we buy some things from UNFI [United Natural Foods, Inc.]." I reminded him that UNFI is a distributor, not a local food producer; that it is based in Iowa; and that is does not carry any locally produced or grown food.
- The Director of Food Purchasing and UM have no policy or program for establishing proper allowances for pricing margins for organic and/or locally produced foods, and they have falsely thrown them into the category of po-

tato chips, soft drinks, and candy bars in terms of retail pricing, economic gain to the university, and need for customer enlightenment.

Now consider that message from President Coleman. The "difference" she urges actually should start with her and with her staff, the administrators, the faculty, and the Director of Food Purchasing. This gross dysfunction is seen not only at UM but also at the majority of our learning institutions across the country. Those at the top, who are making policies and decisions, are disconnected from the reality of what occurs with their food choices. They owe it to their community to establish continuing-education programs and an accurate awareness base for themselves first, before they ask students to "make a difference." The gap between what leaders are saying and the needs of all they serve is filled with layers of lack of enlightenment, irresponsibility, and resistance. Perhaps once they make an effort to understand what "sustainability" and "green" really mean, in terms of our food choices, then change in the right direction can occur.

There is another reason why individuals have not stopped eating meat and adopted a healthier plant-based way of eating. This reason is pervasive and is equally discouraging because many people do not care—they think that they are impervious to the effects of eating animal products. It's the attitudes of "It will never happen to me" and "I won't care until it happens to me." I have witnessed many patients, friends, and relatives who have gone through the typical sequencing of eating unhealthy foods, with a large percentage of those foods being animal products— hamburgers, hot dogs, steaks, pork chops, bacon, chicken, turkey, fish, etc. They have eaten these foods day after day after day, over a period of many years. Then, not so mysteriously, they gain weight and develop one or more diseases—diabetes, cardiovascular disease, kidney or heart disease, or cancer. Eventually, they sustain a life-threatening heart attack or undergo life-changing surgery. They may suffer and die at an earlier age than normal. Many of these individuals feel this occurs because of genetics, which may be true to some extent, but no matter what the genetic predisposition to a certain disease state may be, I can assure you that eating animal products in any form will substantially raise the likelihood that you will contract and suffer from one of these debilitating diseases.

I have even witnessed extreme examples of this with my patients and friends who have eaten meat their entire lives. Some have contracted colon cancer at ages forty-five to fifty and have undergone multiple or extensive surgeries. Yet they then continue to eat meat, even after I presented them with the book *Surviving Cancer* by the Physicians Committee for Responsible Medicine, which cites numerous studies and conclusions that eating meat can and does cause colon cancer, as well as other types of cancer. This, to me, is an excellent example of just how powerful our cultural, social, political, and media influences have been—and obviously still are—regarding the inappropriate perpetuation of the myth that eating meat is good for you. *People are dying because of this.*

Now, let's talk about the physicians in whom we put our trust. A primary reason why people think meat is good for them is because their doctors believe it is healthy and convey that myth. And because doctors are the keepers of our health, we must follow.

Although consumption of animals for food is expected to double over the next two decades, there are stark differences in meat consumption between countries; for example, it's eleven pounds per person per year in India as compared to the United States, which consumes meat at the rate of 270 pounds per person per year.[127] The World Health Organization (WHO). Tufts University researchers, the PCRM, and others have consistently recommended lower intake of animal fat and red meat due to a clear relationship of various diseases (cardiovascular. diabetes. obesity, certain types of cancer) with the consumption of animal products.[128]

Livestock products are also more susceptible to pathogens than other food products and have a capacity to transmit diseases from animals to humans. The World Organization for Animal Health estimates that 60 percent of human pathogens and 75 percent of recent emerging diseases are *zoonotic*—living on or in animals. Many human disease have their origins in animals (such as common influenza and smallpox) and others—such as tuberculosis, brucellosis, and many internal parasitic diseases such as those caused by tapeworm, threadworm, and others—are transmitted through the consumption of animal products. Avian flu, Nipah virus, Creutzfeldt-Jakob disease ("mad cow"). bovine encephalitis, E. coli, salmonella, shigella, Campylobacter. and H1N1 (swine flu) are all associated with handling and consumption of animal products for food.

The wide overuse of antibiotics in animals has caused many bacteria that now affect humans to become antibiotic-resistant. Researchers at Johns Hopkins Bloomberg School of Public Health report that 96 percent of Tyson chicken flesh (Tyson is the largest producer of chickens in the world) is contaminated with antibiotic-resistant Campylobacter bacteria.[129] USDA studies have found that 66 percent of all beef samples were contaminated with bacteria that are resistant to antibiotics.[130] Toxic levels of arsenic are commonly found in chicken flesh.[131] Fish have been found to have levels of PCBs and mercury, thousands of times higher than those in the water in which they live.[132]

USDA inspection reports reveal that on average, one out of eight turkeys served on Thanksgiving is infected with salmonella, and Campylobacter causes the second most commonly reported food-related illness.[133] Reports also showed that more than 50 percent of samples of meat from pigs (pork products) were contaminated with Staphylococcus.[134]

Dairy products contain a wide variety of contaminants, including chemicals and hormones. Milk contains natural hormones and growth factors that are produced within a cow's body and also synthetic hormones, such as recombinant bovine growth hormone (rBGH) or insulin growth factor-1 (IGF-1).[135] Additional contaminants found in milk samples and other dairy products include antibiotics, pesticides, polychlorinated biphenyls (PCBs), and dioxins. When consumed, any of these toxins can build to levels that eventually may harm the immune, reproductive, and other systems, as well as leading to the development of cancer.

A word must be said here about the H1N1 (swine flu) virus and epidemic. On April 30, 2009, the World Health Organization escalated the alert to a level four (out of a possible five), due to worldwide concern for a possible pandemic. Many people died, numerous others were infected, and it spread quickly throughout a number of countries. As it should have been, news of the outbreak and what was being done about it was front and center on every conceivable media format. This is a wonderful example of timeliness and how well information on an important topic can be disseminated in a very short fashion. It is also a perfect example

of *just which information* is really told to us. For instance, with a
story of this magnitude, we know of the first few people who died
from swine flu in the small town of La Gloria in Oaxaca, Mexico.
We know of the generally rapid response of readiness and formal
statements by the United Nations, WHO, President Obama, and
other world leaders. We have been reassured of the stockpiling of
a proper amount of vaccinations, and we even know of certain
organizations' desire to change the name of the virus. I find it
interesting that we have *not* been told how and why the virus exists, which conditions help foster the development of these types
of viruses, and what we should do to remedy the situation. Regardless of what becomes the official statement by investigators
regarding the cause, it will most likely be a clouded version of
the fact that it all began in overcrowded pig farms in that area
of Mexico, which is run by a subsidiary of Smithfield Foods, the
largest pork producer in the world.[136]

Pigs are highly susceptible to both avian and human influenza A viruses; they are commonly referred to as "mixing vessels," in which viruses commingle, swapping genes along the way;
then new strains emerge. It is thought that pigs have been the
intermediate hosts responsible for the last two flu pandemics, in
1957 and 1968.[137] According to the Centers for Disease Control
and Prevention, *up to one-half of pigs on modern farms have evidence of the H1N1 virus.*[138] Thousands of pigs are crowded and
confined in sheds, stacked to the point where the animals are continuously inhaling and recirculating airborne fecal matter, methane, ammonia, and pathogens. Antibiotics are commonly given
to treat and prevent devastating outbreaks within feedlots, but
influenza viruses are resistant to antibiotics. Once a pathogen like
the swine flu virus emerges, it is then spread by farm workers and

by the transport of pigs to other locations.

In the United States alone, over 320,000 pigs are slaughtered for food *every day*, which drives the continual operation of congested livestock farm lots.[139] In the Mexico town near the Veracruz Mountains, where this recent outbreak began, more than 450 residents had complained of severe respiratory and flu symptoms weeks before the outbreak and confirmed swine flu virus strain, which affected and eventually killed a four-year-old boy there.[140] The focus of all our attention, therefore, should not be on which schools to close, when to wear masks, who should be vaccinated, who should be allowed to travel Mexico, or what to call the virus. The majority of our efforts should be on divulging the real reason behind this epidemic, which is the factory farming of massive amounts of pigs in filthy, confined conditions that promote the development of viruses that can cause infectious diseases in humans. And the reason this happened is because of our demand for meat products. Those pigs are here, living in those conditions and developing viruses, only because people want to eat them. Therefore, I find it incongruous that there has been such a movement to remove the name "swine," and major efforts have been made to assure the public that this virus has nothing to do with pork products and that they are entirely safe to continue eating. Well, the virus has everything to do with pork products. The movement to remove the name swine is propelled by the USDA and world pork producers, and although it is true that the H1N1 virus may not be contracted directly by eating pork, the production of pork is precisely the ultimate reason that the swine flu virus exists. So while the pork industry is encouraging people to continue eating pigs, it is the eating of pigs that is the problem—this is exactly what the public needs to be told so that the problem can be resolved.

This completes the portrait of what I consider to be global depletion of our own health. Food choices are implicated, and you have the ability to change that.

CHAPTER IX

Tread Lightly

Entering the no-wake zone

"The greatest obstacle to discovery is not ignorance;
it is the illusion of knowledge."
—Daniel J. Boorstin

GLOBAL DEPLETION IN SOME FORM

will occur simply because the earth can only support so many people doing so many things over so long of a period of time. The gross number of humans on our planet is not as much of an issue as what they are all doing to the planet in a short period of time. To make matters worse, individuals and institutions that are in a position to expose myths, enlighten the public, and change the direction of public opinion clearly are not doing so. There are two primary reasons for this: first, lack of adequate knowledge, although they purport to have it; and second, they are unable to express the truth due to various constraints—political, cultural, social, legal, business, etc. Although these entities may be accomplished in some particular field, it essentially has provided them an avenue to influence us in other areas. They may not be in these elevated positions because they are any more brilliant than the rest of us or have any special skills that others do not have. While we must respect that talent or knowledge play a role in the attainment of an elevated status, they also have access to a platform for a variety of other reasons, with an accompanying form of media that allows them to express their opinions and influence their large audience. Sometimes this is a good thing, but more times than not, there is lack of full disclosure, even an ulterior motive. This is one of the reasons that full enlightenment of an important subject frequently does not occur. There are many examples of this, but some that immediately come to mind are with newscasters, best-selling book authors, actors, and especially hosts of gossip shows, politicians, and organizations/occupations that we hold in such high esteem, such as doctors and dieticians, specific businesses, and institutions. Unfortunately then, this can compose up to 99 percent of our current mode of information. Some talk show hosts, such as Oprah Winfrey, are placed in such high reverence that we will support anything they recommend, whether it is a new book, movie, or presidential candidate. There is one caveat, however—whatever these people with platforms have to say, it cannot be alarmingly controversial, especially as it relates to food. If any comments are made that would negatively impact the meat, dairy, or fishing industries or our current cultural dependence on these, a career could be jeopardized. Therefore, these high-profile people must tread lightly and not create waves.

Over the past forty years, certain individuals have enlightened themselves by researching the general topic of the detrimental effects of eating animals and have essentially arrived at similar findings—that our demand for consumption of animals for food is not healthy or sustainable. Most of them have arrived at this conclusion by close examination of one aspect of the problem, either human health, pollution, land use, crop and feed use, water supply, animal rights, or something similar. For instance, thirty years before the American Dietetic Association even acknowledged the health benefits of a vegetarian diet, Nathan Pritikin and others wrote about epidemiologic studies that exposed the clear relationship between eating animals and the development of numerous Western diseases. Twenty-five years before the United Nations published a paper regarding our current unsustainable use of livestock, many researchers and authors, such as Jeremy Rifkin in his book *Beyond Beef*, explained the many problems that the cattle industry has caused. These individuals did not tread so lightly.

So if this information has been available, why haven't you heard about it? And why isn't anything being done now to move people in the right direction of choosing foods that will not kill us or our planet? The reason lies on three levels:

1. Individuals who know this information are not given an adequate platform to get the message out. Essentially, the information never goes anywhere.

2. There is a disparity and absence of knowledge with highly visible individuals who have been afforded speaking platforms and otherwise have the ability to inform and influence the public. Here, the information could certainly be disseminated, but these public speakers, celebrities,

or "experts" are unaware and do not have all the correct information about food choices. The public is essentially influenced by inadequate or incomplete data and misinterpretation provided for them.

3. People who are aware and are in various positions to get the message out so that it could make a difference do not speak about it. (These individuals are usually in the media, such as news commentators, prominent actors, or talk show hosts.) Why? They may be afraid of the potential repercussions in providing information that is controversial and which might be difficult for our culture to accept. These individuals also may be policy-makers, such as legislators, who come across this information somewhere but would feel uncomfortable taking the appropriate action. Although it would save millions of lives, reduce health-care costs, and save our planet, and even though they really may want to inform the masses and make a change, they realize it might be a bad move for their job security—numerous powerful businesses and industries would have difficulty with this knowledge becoming commonplace.

Nowhere is the "tread lightly" concept more evident than with Oprah Winfrey. On April 16, 1996, Oprah allowed a discussion on her show that divulged accurate accounts of unsafe feeding of livestock and subsequent disease outbreaks, as well as eventual illnesses and deaths of consumers. When Oprah declared she would "stop eating hamburgers because of fear of mad cow disease," the powerful National Cattlemen's Beef Association (NCBA) retaliated by taking her to court. This caused difficulty for her, as she had to move her show to Texas while she was caught in the defense of her case. Since then, there has been no discussion of any topic that could even remotely be considered negative regarding the meat and/or dairy industry. She now says only what is politically appropriate for continued viewer support.

It is interesting to note that this topic of contracting food-borne illness from eating beef, which created such a furor to the NCBA, was simply one of many topics that could have and should have been discussed to enlighten the public. There is so much more room for talk show hosts to allow elucidation on the negative impact our demand to eat animals has on the environment, our health, health-care costs, or a variety of other related subjects. I consider it fortunate for the NCBA that Oprah exposed to her audience only one small fraction of the multitude of serious problems for which the meat industry is directly responsible.

A classic example of misuse of a media platform with regard to eating meat was seen on *The O'Reilly Factor*, hosted by Bill O'Reilly. On Thursday, January 29, 2009, O'Reilly had an ideal opportunity to educate his audience to a better understanding of the negative aspects of eating meat during the segment titled "Is Meat a Good Idea?" Rather than doing his homework on the subject, however, and choosing appropriate guests and allowing obviously important but previously suppressed issues to be discussed intelligently, Mr. O'Reilly allowed continued misperceptions on the topic to grow. Somehow, the segment "Is Meat a Good Idea?" was turned into a unilateral and scientifically unsupported discussion about erectile dysfunction. This had almost nothing to do with the topic title. Then, equally off target, the discussion turned to PETA (People for the Ethical Treatment of

Animals), and ended with O'Reilly stating, "I eat meat and resent being told by PETA that I'm some kind of savage for doing it." In actuality, Mr. O'Reilly may not be a "savage," but he is quite uninformed and in a state of denial regarding his continued unhealthy choice of foods and the misuse of the informational platform he commands. This could have been a wonderful opportunity to reveal facts to the viewing audience and to increase their understanding of the ill effects of eating meat on their own health and that of our planet.

In *An Inconvenient Truth*, Al Gore enlightens readers with information on global warming and suggestions for lessening our carbon footprint. This was a wonderful thing, and it earned him a Nobel Peace Prize. As you now know, however, Gore discussed only part of the story. The much larger story is that of the livestock industry's role in producing more global warming than all our cars, trucks, planes, and other vehicles used in transportation combined. What is interesting to me is that Gore epitomizes the "tread lightly" issue, as he was quite aware of this information. He chose to seek the path of least resistance, the path to tell essentially the convenient truth and mention, in an obscure area at the back of his book, that we should "modify [our] diet to include less meat."

His *An Inconvenient Truth* is filled with examples of how the earth is changing as a result of global warming, but it does not provide a connection for the reader to the real culprit. He addresses the theme of logging, stating, "The way we treat forests is a political issue." He fails to mention, however, that the *reason* for deforestation, or "logging," is because what we choose to eat accounts for over 70 percent of all forest lost in the Amazon region. He had ample opportunity to provide readers with accurate reasons, but he chose to call it "logging."

I can only speculate on the real reason for Al Gore's non-disclosure, which I suspect are his own ties to the current livestock industry. He owns a farm and has cultural and perhaps political affiliations with that industry, and it would be a controversial and risky move for him to give too much attention to the adverse affects of that industry. I appreciate his efforts and accomplishments with one aspect of awareness to global warming, but he did not appropriately use the public platform he was provided to present the correct message, one that would be the most effective to promote change for a healthier planet.

As we saw earlier, the Kyoto Protocol was agreed in December 1997, although it finally took effect on February 16, 2005. While all countries that joined under the Convention are "encouraged" to stabilize their greenhouse gas emissions, certain nations who are signatories to the Protocol are legally committed to an average reduction of 5 percent of 1990 emission levels. These reductions must occur between 2008 and 2012. As of late 2009, 187 parties had ratified the Protocol, but there was no replacement framework to follow from the year 2012. Because of this, a fifteenth session was held in December 2009, the Copenhagen Accord, with the intent to create a framework as a follow-up of the Kyoto Protocol to "define methods of reducing emissions and how to offset what we cannot" (UNFCCC, 2010). Results were that "pledges were communicated by 75 parties to cut or limit emissions of greenhouse gases by 2020." Pledges are a step—although clearly a baby step. Because again, nowhere in the reports of the UNFCCC is there specific wording that addresses the fact that the livestock industry is a major cause of global warming, that it is driven by our demand to eat animals, and that strategies, therefore, need to be developed to reduce and eventually eliminate this factor.

I know of one solution that perfectly fits both categories of strategies; that is, how to reduce greenhouse gas emissions and how to create offsets: simply stop raising livestock. This is, of course, a large step, but it is the right step. And doing so would substantially reduce greenhouse gas emissions, and at the same time, major offsets would be created by the concomitant restoration of forests and other terrestrial, as well as oceanic natural habitats.

Mark Bittman's book *Food Matters: A Guide to Conscious Eating*, also had the opportunity to provide enlightening information and dietary solutions related to meat and its large carbon imprint. Instead, Bittman used it as a twist to sell another book. Although on the back cover, he uses phrases such as "help stop global warming" and "requires no sacrifice," he still advocates eating meat and animal products throughout his book—warm bacon dressing, squid or shrimp, turkey thighs, pork or lamb shoulder, beef chuck, Italian sausage, bone-in pork chop, comfit duck legs, and the dish he "craves all the time ... Thai beef salad with flank steak." How does eating these foods help to stop global warming? How does eating these foods "reduce your risk of chronic disease" and require "no sacrifice," as he emphasizes in bold red letters on the back cover of his best-selling book? In fact, eating the foods he advocates has exactly the opposite effect. Those foods increase your risk of many chronic disease states, increase global warming, add greatly to global depletion, and certainly require a heavy sacrifice of resources used, as well as the animal itself that you are eating.

Mr. Bittman's book can be more easily accepted by the

public and likely will put him on television and radio shows, and provide exposure for marketing purposes. The result of his not addressing the true facts, however, is that his book directly implies that it is okay to continue eating meat; that it's okay to continue choosing foods that cause significant global depletion and health risks to ourselves.Clearly, it is not okay at all. Individuals such as Mr. Bittman, who are provided with a platform, should provide the correct information or give the platform to someone who is more capable and concerned about creating proper direction.

The most profound example of the politics of food systems and corporate influence on you can be found in *Food Politics*, a book in which author Marion Nestle divulges the vast grip the meat and dairy industry has on the U.S. public and the way in which they limit and affect our food choices. In 1986, Dr. Nestle moved to Washington DC to work for the U.S. Public Health Service, where she managed the editorial production of the first government-directed book about diet and health in America, the seven-hundred-page "The Surgeon General's Report on Nutrition and Health." It can be explained no better than in her introduction: "My first day on the job, I was given the rules: No matter what the research indicated, the report could not recommend 'eat less meat' ... or the report will never be published." Research, in fact, did undeniably indicate that meat and dairy products were linked to various disease states and that eating animal products of any kind—but particularly meat and dairy—would substantially increase our risk of heart and cardiovascular disease, hypertension, gall and kidney stone development, and some cancers. Instead of being able to report these findings and provide appropriate suggestions to the public in a straightforward manner, Dr. Nestle was coerced into incorporating them in a suppressed fashion that would not be so detrimental to the powerful livestock industry. Specifically, instead of stating the facts of these findings and that consumption of meat and dairy should be substantially limited or eliminated from the diet, the book uses wording such as "choose a diet low in ..." or "have two or three servings of meat." This effectively allowed the meat and dairy industry to continue the cultural brainwashing of the American public through misleading and misinforming marketing.

The case against consuming livestock could not be more obvious than in the 2006 LEAD report by Steinfeld et al, and yet there is a blatant display of treading lightly revealed by the conclusions. The Livestock, Environment, and Development (LEAD) initiative was formed to "address the environmental consequences of livestock production, particularly in light of the rising demand for food products of animal origin and the increasing pressure on natural resources." The LEAD initiative was supported by the United Nations, the World Bank, the European Union, and numerous other international organizations. As stated in their executive summary: "The livestock sector emerges as one of the top two or three most significant contributors to the most serious environmental problems, at every scale from local to global." Findings suggest there should be major policy focus dealing with livestock problems "of land degradation, climate change and air pollution, water shortage and water pollution, and loss of biodiversity." Knowing now that livestock's contribution to environmental problems is on such an enormous scale, they further conclude: "The impact is so significant that it needs to be addressed with urgency." The solution does seem pretty obvious: stop eating meat. However, instead of simply advising lessening the demand for livestock by reducing the consumption of animals and animal products, the LEAD authors provide suggestions such as "relocating factory farms away from urban areas," "finding feed that results in less methane in the animals' flatulence," and others that essentially perpetuate the problem, both conceptually and functionally. These authors, who spent so much time and energy uncovering the fact that essentially all aspects of the livestock industry are devastating to our planet, summarize by saying that even though we are losing our planet on many levels due to livestock, let's keep raising them and eating them—it's okay to do this as long as we find ways to make the devastation less profound.

We are hitting ourselves in the head with a hammer repeatedly, on the way to killing ourselves, such that scientists who have studied this problem exhaustively suggest we put a Band-Aid over the head wound but keep hitting our head with the hammer. This is the ultimate in "treading lightly," rather than concluding the obvious: just stop hitting yourself and take the stupid hammer away once and for all.

Another form of treading lightly also can be found with contaminated food outbreaks. Each year there are thousands of reported sicknesses and hundreds of deaths due to food-borne pathogens such as E. coli, salmonella, shigella, and Campylobacter. The outbreaks are always reported in regard to which food source it was associated with. Typically, it is meat products, which is far under-reported or minimized. But there are many cases that involve plant-based foods, such as melons, tomatoes, salad mix, spinach, peanuts, and pistachios. So, how does treading lightly fit in? Well, in 100 percent of the reports by the media over the past twenty years, none revealed the real source and, therefore, the reason for the outbreak. In all these reports, the illnesses were blamed on the specific foods. The vital piece of information that is routinely omitted is that all these pathogens that cause sickness are actually caused by animal sources. For instance, salmonella grows abundantly on chickens and other animals, so when they are killed and eaten for food, there is a high likelihood that salmonella may find its way to humans. This similarly occurs with E. coli, which is found in all animals. It should be no surprise that because it is found in all animals, there is a probability it will show up somewhere down the line if you eat those animals. What is never discussed is that nowhere on any plant-based food can these same pathogens be found naturally. Yet investigators fail to mention this important point.

Then what causes salmonella, E. coli, and other food-borne diseases in vegetables, fruit, nuts, and grain? Plants can only be contaminated by coming into contact with polluted water through irrigation, animal fertilizers, and using animal or human feces. Vegetables and fruit can also become contaminated if placed in close proximity to or mixed with raw poultry, meat, or eggs, and unpasteurized milk, as all of these products have supply bacteria contaminants on them naturally. When E. coli was found in salad mix and spinach, it was never mentioned that it was found in these products because they were irrigated with water that was contaminated from a cattle operation a few miles away. The cattle business allowed its manure to be washed into a water system that eventually made its way to surrounding vegetable farms.

Investigation of a rather large outbreak of salmonella in peanuts recently revealed that the cause was "unsanitary conditions at the plant."[14] FDA inspectors found at two Peanut Corporation of America plants at least four strains of salmonella, all caused by "rodents, bird feathers, and rodent excrement."[142] Again, the strains of E. coli and salmonella that caused these outbreaks occurred due to contamination from animal sources. Animal products used for food have a high propensity for distribution of pathogens, because they are found naturally on and within the animal itself during its life, as well as during the slaughtering process. This applies to all animal products, whether it is from livestock, dairy, or fish. This adds to the list of ways that eating animals creates a depletion of our health—and are examples of treading lightly.

In mid-August 2010, more than a half-billion eggs were recalled from numerous areas in the United States due to a salmonella outbreak. Thousands of illnesses resulted from this contamination, which causes abdominal cramps, diarrhea, fever, and sometimes death. While news of the outbreak and recall spread quickly throughout various media sources, there was a consistent omission of providing the public with a fail-safe solution. One example was broadcast on National Public Radio on August 25, 2010, during an interview by correspondent Linda Wertheimer. This particular interview was with Dr. William Schaffner, chairman of preventive medicine at Vanderbilt University, who provided perfect answers to most of Ms. Wertheimer's initial questions regarding how eggs become contaminated in the first place. He informed listeners that salmonella is found living normally in all hens, primarily within the intestinal tract, and therefore it is common to find this pathogen on hens, on their eggshells, and in the interior of eggs. The difficulty with this interview was in the information given to the listener in terms of solutions. There was ample discussion from Ms. Wertheimer and Dr. Schaffner about the use of eggs derived from free-range chickens, as well as cooking and cleaning techniques and use of pasteurization to kill the pathogens. Nowhere, however, was there mention of simply not eating eggs. Both participants had a perfect opportunity to suggest that we *not* use eggs—or chickens, for that matter—as a food source. After all, that is a possible solution, isn't it? And it's a solution that would actually eliminate the problem entirely as it relates to eggs, and at the same time reduce our inefficient use of resources, such as land, water, and food. It seems easy enough to have posed the question or to have offered it as a solution. Instead, both Ms. Wertheimer and Dr. Schaffner—as well as, perhaps, the writers and editors of this NPR segment—chose to ignore this very obvious choice, essentially treading lightly with this more controversial topic. It is unfortunate that the estimated 21 million NPR listeners were left with half the story and no real resolution.

The "tread lightly" phenomenon is vividly displayed by author Michael Pollan, who is known as "America's most trusted voice on diet." While Mr. Pollan has brought some focus and even validity to eating more of a plant-based diet, his primary agenda is to bring attention to the "industrialization" of our food and the ill effects it has on our health. His books are eloquently written, and he presents concepts with intelligent conviction. He has uncovered more than his fair share of facts related to all the detrimental effects of eating any animal products—and the evidence is exhaustive. Whenever he is asked if he eats meat, he repeatedly goes on the record as saying, unequivocally, yes. He also promotes our use of grass-fed livestock and creates the false illusion that it is healthy and fully sustainable. Neither point could be farther from the truth. Providing pasture-fed cows, pigs, chickens, turkeys, or any other animal is fully *unsustainable* and unhealthy for our health and the health of our planet. I believe that Michael Pollan is "treading lightly" so as to not diminish his audience. It is so much easier for us to identify with his message if he says that he still eats meat, like the rest of the world. Perhaps, however, even with all his research, the obvious has not been absorbed. Perhaps Michael Pollan is *comfortably unaware.*

CHAPTER X

How We Arrived at This Point

Observations, predictions, and solutions

"We are made wise not by the recollection of our past
but by the responsibility of our future"
—George Bernard Shaw

ALTHOUGH GLOBAL DEPLETION

has occurred in many areas and in a devastating manner, there is hope—but certain things must change. Addressing complex issues such as our food choices must immediately happen on many levels. There must be enhanced awareness and vision, proper prioritization, subsequent implementation, and dissemination of truth to everyone involved as progress occurs. Before we can chart a better course, we should first examine how we arrived at this point ... the path we took to get here.

Collectively, we choose food that is unhealthy for us and for our planet. There are a number of reasons for this unfortunate situation, including:

- Clear suppression of information
- Misplaced trust for guidance
- Misleading food choice education by the USDA and the meat and dairy industries
- Government subsidies for animal products
- Lack of establishment and implementation of an ecotax, reflecting a true price
- Complex combination of psychological, cultural, social, and political interactions

There is very little awareness that your food choices have a profoundly detrimental effect on our planet. However, it should be common knowledge that, much like smoking cigarettes, eating meat is not healthy for you. Strangely, consumption continues, so there must be something missing in this scenario. Let's look once again at what a few of the major health organizations say about the consumption of meat and dairy products and the effect it has on your health:

- The American Dietetic Association states: "It is the position of the ADA that vegetarian diets, including total vegetarian or vegan diets, provide health benefits in the prevention and treatment of certain diseases."[143]
- The American Institute for Cancer Research and the World Cancer Research Fund call for "a plant-based diet rich in a variety of vegetables, fruits, and legumes and

limiting red meat consumption, if red meat is eaten at all."[144] [145]

- The American Heart Association and the National Institutes of Health call for a diet based on a variety of plant foods, including grain products, vegetables, and fruits to reduce risk of major chronic diseases.[116]

These are all purposeful, visible organizations with a clear message regarding food choices, so this begs the question: Why do people continue to eat meat? My experience has shown that individuals will fall into one of the following categories:

- **Burger, Fries, and a Side of Disease:** You are quite aware of this information and have made a decision to significantly increase the probability of developing cancer, heart disease, diverticulosis, gallstones, kidney disease, and diabetes, and will spend most of your time in physician offices and hospitals, and buying medications, like the majority of Americans do.
- **I am Superman:** You are aware of this information and the connection but believe it will never happen to you, even though statistics show it happens to the vast majority of everyone who consumes animal products.
- **Need My Protein:** You really are completely unaware that there is anything wrong with eating meat and may think it is actually healthy for you. If you are in this group, you are part of a large number of people who are hearing, seeing, and reading only what the various aspects of our culture want you to know, in order to perpetuate the belief that those products are healthy.

This third group exists due to a rather complex intertwining of cultural, political, economic, and media factors that, when combined, does an excellent job of not allowing the average person to become aware of pertinent facts. Even if you had these enlightening facts in hand, there is then difficulty in feeling comfortable with taking the proper course of action, due to certain social and cultural implications. Unfortunately, this is the situation that affects the majority of Americans. And therein lies the problem.

For instance, we respect, almost to the point of reverence, our physicians, dieticians, and hospitals—the keepers of our health. Because of this, we have come to rely solely on them for guidance when we become ill or injured. This is perfectly fine in certain circumstances, but it also places these professionals in a counseling position regarding disease prevention and nutrition. Now, this is where problems occur.

First is the dichotomy of their education and cultural influences, relative to the position of guidance in which we place them. Physicians are trained in medicine; they are medical doctors (MDs). As such, they are trained in diagnosing and treating various disease states, primarily with a pharmacological solution—which drug will manage which condition. The primary focus of their education was not in prevention, nutrition, or diet and exercise counseling, nor was it in ecology issues, to be able to determine what is in the best interest of our planet. Medical schools do not require their students to take nutrition courses beyond one semester—and it is an elective in most schools. It is quite easy to see why most Americans feel it is healthy to eat meat and animal products in general, *because their physicians mistakenly think that it is fine.*

Today, it is standard procedure for heart surgeons and cardiologists to prescribe medications and perform bypass surgery without mentioning that prevention or treatment could be best managed by switching to a completely plant-based diet. This happens despite current knowledge that supports this recommendation. Similarly, those suffering from colon, rectal, pancreatic, or prostate cancers are treated with extensive, life-changing surgeries, chemotherapy, and radiation. Oncologists and surgeons seldom mandate a plant-based diet. I have witnessed countless patients, immediately following massive surgeries to remove cancer of the colon, who are told they can eat whatever they would like. There is no mention of statistics or current medical knowledge that eating animal products most likely contributed to the development of the cancer; no mention that eating animal products will significantly increase the risk that more cancer will develop. (This applies to many other types of cancer as well.)

Dieticians are equally, if not more, at fault. As hospitals become more occupationally specialized with those in decision-making capacities, dieticians play a much larger role than they did even twenty years ago. Hospital administrative personnel place dieticians in supervisory roles, with meal planning for the entire hospital. Physicians rely on dieticians to formulate meal strategies and guide pre- and post-surgical patients on food choices that are in their best interest. Unfortunately, nearly 100 percent of the dieticians I have come in contact with over the past thirty-five years either have no knowledge of the health benefits of a plant-based diet or are still in denial that meat and animal products are factually unhealthy for anyone to consume.

Since 2003, the year in which the American Dietetic Association printed its position statement that highlighted the many advantages of a vegetarian diet, I have yet to meet even one dietician who uses this information as a basis for meal planning for her patients, despite its obviously being in the patients' best interest. Maybe dieticians have not read their own position statement, which was reaffirmed and reprinted in 2009. More than once, I have visited a patient, one or two days out of surgery in which the colon and rectum were removed, and witnessed the physician and dietician allowing or even directing the patient to eat hot dogs. It is easy to see why we continue to think that eating meat is healthy.

I personally know many pediatricians who continue to provide the nutritional guidance that milk and dairy products are healthy—I am sure this applies to the vast majority of pediatricians. Not only do they allow their young patients to consume milk, but they most likely promote its use for their own family members. This occurs despite the fact that the American Pediatric Association has recently stated that milk and dairy products should not be given to children, as they are a recognized source of allergens and contaminants, such as hormones, antibiotics, and pesticides. Dairy products also cause colic and chronic constipation. Additionally, some studies have now linked milk consumption to children's developing type I diabetes.[117]

With health professionals continuing to advocate the consumption of meat and dairy products, true wellness for us and for our planet will be impossible to achieve. This brings light to the fact that we have a system of suppressed information and a misplaced trust in individuals and institutions that provide inaccurate and misleading information. All this needs to change in order for progress to occur.

The mismanagement of information or outright ignorance of reality regarding our food choices and the toll that it takes on our environment is the reason why we are at the point we are today. Still, one simple illustration has yet to be made: have you actually raised a cow on your own property? Not too many of us can say that they have. If you have and really understood what you were doing and your full spectrum of options for food, I do not believe you would ever continue the practice. Why? Because that cow would drink forty times more water than you do, eat thousands of dollars worth of food that could have been used in one form or another for you, use all your land, and create urine, feces, and flatulence that is overwhelming. Then there is the killing process, which will consume additional water and energy, just to get a few parts of the cow to your table to eat, which is not healthy for you, despite the effort. Now, think about this on a scale of 70 *billion* animals in one year alone.

Another large reason for our current state is because of the education process imposed on us by the meat and dairy industry, with their massive multibillion-dollar advertising campaigns, lobbying, and political efforts. They infuse us with thoughts that their food is healthy for us, and of course, there is no mention of what they are doing behind the scenes to our planet. For every minute that we are exposed to a purported health benefit of an animal product, we should be equally exposed to the ill effects those same products have on us and our planet—how much water, land, energy, and resources it took and the pollution created to get it to you. And we should be equally enlightened to all the benefits of a plant-based diet for our health as well.

Misleading advertisements abound. The origin of our Food Pyramid was nearly a hundred years ago, with the USDA Food Guidelines in 1916, which promoted meat and dairy products at the center of every meal, in order to obtain the proper nutrition.[148] These guidelines and subsequent Food Pyramid served—and still serve—as the Holy Grail for our country's school systems and families with regard to food choices. Influenced heavily by the egg, milk, and meat producers, the USDA has misdirected the public without much challenge for all these years, by falsely promoting animal products as being necessary for good health. The USDA and Food Pyramid still promote meat and dairy items, with dairy as high as "five servings needed daily for teenagers and pregnant women," despite the fact that milk and meat products are the largest sources of saturated fat and cholesterol in children's diets, according to the National Institute of Child Health and Development.[149]

The USDA and its guidelines influence much of the world in terms of food choices, and it still dictates what 17 million food-stamp recipients can eat and what foods can be offered through the National School Lunch and Breakfast Program. No wonder animal products continue to be consumed; no wonder we are in the worst overall global health we have ever been in. A newer Food Pyramid, with suggested revisions established by the Physicians Committee for Responsible Medicine, places animal products, such as meat, dairy, and fish in the proper location—off the pyramid entirely.[150] Even so, the Dairy Council continues to heavily advertise with campaigns such as the milk mustache worn by recognized celebrities and announcing that milk is good for strong bones and preventing osteoporosis. Now, they even have one that declares milk will help you lose weight. Again, nowhere is there equal information for the consumer regarding the evidence that milk promotes various disease states, has unhealthy contaminants, is devoid of certain important nutrients, and that there is a heavy burden on our environment with every drop created.[151]

Another reason we are at this point with our food choices is that there are subsidies paid to the meat, dairy, and fishing industries to help support them, regardless of demand. This economic assistance is quite disparate in the sense that there are no subsidies of a similar nature for plants grown directly for our consumption. So, although there are really no logical arguments for the continued use of animal products as food, our government has created a shield of sorts for the meat and dairy industries, particularly as it relates to normal economic or market fluctuations. In 1933, the USDA created the Commodities Credit Corporation (CCC) and began giving direct price supports to dairy production and de facto supports to the meat industry in the form of feed grain price assistance.[152] This has allowed our government to keep the industry in an artificial sense of security and viability and immune to any downturns due to market pricing or demand. For instance, in 1998 USDA Secretary Dan Glickman bought up at least $250 million worth of beef, chicken, dairy, eggs, fish, lamb, and pork that could not be sold in an already flooded market.[153] This kept an artificial money supply flowing back to the meat and dairy industry that helped to perpetuate production of their animal products without true market or public demand. These goods were destined to be dumped into public feeding troughs such as the National School Lunch Program.[154]

Subsidies are even created for the livestock industry for the use of water that promotes overuse of our water supply and without proper taxation. Water use has been cost-free with western cattle ranches that enjoyed access to streams and rivers that have now dried up altogether, due to overgrazing, soil erosion, desertification, and general overuse. The government has essentially encouraged pumping more and more water from underground aquifers, as well as further freshwater depletion, by creating tax deductions for sinking wells and purchasing drilling equipment. It has sponsored more than thirty-two irrigation projects in seventeen western states. In all, more than half the cost of providing irrigation facilities in the United States has been borne by the federal government, which has subsidized ranchers and farmers from public funds.[155] As stated by Cornell University economist David Fields: "Reports by the Water Resources Council, Rand Corporation, and the General Accounting Office made it clear that irrigation water subsidies to livestock producers are economically counterproductive ... current water use practices now threaten to undermine the economies of every state in the region."[156]

One of the primary reasons we are seeing depletion of our oceans and its marine life is the irrational subsidies on a global basis that is given to the fishing industry. The global fishing industry is now receiving an estimated $34 billion annually, in the form of financial assistance, marketing support, modernization programs, storage infrastructure improvements, boat construction, and buy-back incentives, foreign access agreements, and massive tax exemptions.[157] Total world subsidies for fuel alone are currently at $6 billion, which continues to support the industry's overfishing methods.[158] As stated recently by the World Trade Organization, "Eliminating capacity-enhancing fisheries subsidies is the largest single action that can be taken to address global overfishing.[159] These subsidies create strong economic incentives to overfish." As previously noted, 70 percent of the world's fish species are either fully exploited or depleted, with areas of the ocean having become essentially wastelands and with ecosystems lost. Fishing fleets worldwide are estimated at 250 times more than what is needed to carry out sustainable fishing.[160] Government subsidies have been linked to illegal, unreported, and unregulated fishing and provide the support for large, distant water fleets to fish around the world.[161] Analysis of high-seas trawling, which the United Nations has called to be significantly restricted due to its destructiveness, has found that these fleets would not be profitable without the large subsidies they are handed. With elimination of all subsidies given to the global fishing industry, in combination with less demand by the consumer, the largest component of unnecessary destruction and depletion of our oceans would be eliminated.

There are so many more examples of this, where the meat, fishing, and dairy industries put pressure on the government to continue price supports and purchase programs, while false profit margins are used to fuel their massive advertisement campaigns, nutritional education of the general public and school systems, and political lobbying actions to ensure that benefits continue. This is an insidious cycle, which obviously needs to be broken to allow proper development of a national health and nutrition program, whereby growers of nutritious and sustainable organic food products, such as grains, fruits, and vegetables, are assisted for their efforts. Accompanying this, educational programs can then be implemented so that proper and accurate nutrition information is disseminated to the public and the school systems.

Importantly, another reason that we are at this serious point with global depletion from our unhealthy food choices is the issue of proper pricing. To date, there has been no accountability for a reflection of the true price of a particular food item, to get it from point A to point B. While I have addressed the clear global depletion that occurs with our choices of various animal products used for food, none of them carries the true cost of producing that item. What do I mean by "true cost"? Let's take a closer look at that quarter-pound burger you bought for lunch.

You may have paid $3.39 for it. The true cost of the bun, condiments, and plant portions may have been around $0.30, and the quarter-pound of hamburger, as it is called, has a true cost of … what? It requires fifty-five square feet of rainforest to produce that much meat, so what is the cost of the rainforest loss and all the vegetation, oxygen, and carbon dioxide disruption, as well as the biodiversity lost with it, and why wasn't that accounted for in the $3.39? In many cases, it requires over 1,200 gallons of water to produce just one-quarter pound of edible muscle tissue on a cow. If that water came from a source such as the Ogallala aquifer, which much of your meat does, it will never be replaced in our lifetime, so what is the real cost of that 1,200 gallons that you just used? These examples and questions raise the all-too-important issue that our environment and natural resources have been used as if there is *no* cost, when there obviously is.

As we have seen, there is irreversible depletion with the production of the meat you just chose to eat. All of the resources on earth that comprise our various ecosystems and environment belong to earth and its inhabitants, collectively. As such, it is our duty to act as stewards, leaving the earth in similar or better shape when we pass it on to the next generation. Therefore, our resources should be viewed in a less narrow-minded perspective. These resources are not free and should have some form of an eco-tax affixed. If such a tax existed, then it would affect your burger purchase in the following ways:

There would be a true cost established, whereby the retailer (in this case, McDonald's, Burger King, Wendy's, etc.) and producers would have to pay for the resources that were used in producing that burger—pay for the water used, pay for the rainforest destroyed, pay for the oxygen lost by the vegetation cut down or used, pay for the carbon dioxide it produced in the vegetation loss, and pay for the global-warming gases (methane, nitrous oxide, and carbon dioxide) produced in the raising and slaughtering of the cow needed. Because some of these things are irreplaceable in our lifetime, it would be difficult to estimate the true ecotax, but somewhere in the thousands of dollars would not be out of line for replacing a five-hundred-year-old section of Amazonian rainforest.

Once this ecotax is established and implemented, the obvious effect would be that the only thing that meat and dairy industries could afford to produce—or that you could afford to eat—would be the plant-based portion (lettuce, tomato, condiments, pickle, bun) of that quarter-pounder sandwich. This would set into motion the quickest recovery route to a healthy planet.

Finally, a very complex intertwining of psychological, social, political, and economic issues has been created by the factors mentioned above. These issues affect most individuals as they make food choices on a daily basis. As discussed, there are political and economic issues at play with corporate forces that affect policies, distortion of public information, and even pricing of food. From a psychological and social perspective, it is generally very difficult for most individuals to eat only plant-based foods, due to lack of availability and various constraints placed by parents and friends and within social groups. It becomes just too much of a hassle to ask your friend, colleague, or waiter to not have any animal products in or near the food you are eating. Also, it cannot be cooked in the same pan or touch the same area that meat, fish, or dairy has touched, and so on. Although quite true and consistent, you suddenly become the outcast.

Now that we have a better understanding of which factors led us to the point where we are today, it should be easier to identify calculated predictions as well as further exploring solutions. Production of meat and dairy is expected to double in the next ten years, rising from 229 million tons to 465 million tons.[162] Our planet simply cannot sustain this because it cannot sustain our current levels of raising animals for food. There are two principal ways we can move forward with our food choices regarding animal products and global depletion. First, we can evolve to the point of not eating them by correcting the issues presented earlier in this chapter. Second, we can move forward with legal coercion, because, most likely, the first avenue is idealistic and with the potential of excessive delay. We will reach a point in time where legislation will be forced to enact sanctions that make it illegal to eat meat more than once a week and then, ultimately, at all. This may seem a radical thought at first, but when we reach the point of urgency, it will be one of the necessary corrective measures.

The end result is necessary to halt the global depletion that is currently out of control with our food choices. I call it the "K-Pax" theory. In the 2001 movie *K-Pax*, Kevin Spacey plays a visitor from another planet, one that is vastly more developed and advanced than Earth. There are many subtle references throughout the movie that indicate that his character and all individuals on his planet are vegetarians and do not eat any animal products. It is clearly implied that as a civilization evolves, it must become vegetarian in order to survive and become advanced. Although this was simply a movie, it is intellectually on target, which is why Einstein pondered the thought frequently and arrived at the very same conclusion: "Nothing will benefit human health and increase the chances for survival of life on earth as much as the evolution to a vegetarian diet."

The first way of change through evolution involves correction of all the reasons of why we are here to begin with. Opening the pathways of communication regarding the reality of food choices would be a start. Those who have platforms in the media—celebrities, authors, talk show hosts, and politicians—all must assume a higher level of awareness and then convey the realities to their audiences. Information regarding the benefits of a plant-based diet to our health and that of our planet must be made available to everyone and repeated on a daily basis in the media. Likewise, the ill effects to our health of consuming animal products and the devastating effects it has on our environment must be made available to the public and reinforced just as often. Only accurate, unbiased information must be free-flowing and disseminated to everyone.

There must be a balancing or equating of the reliance we place on health professionals with their level of knowledge regarding our food choices. In other words, if physicians or dieticians are to remain in a position of counseling other humans as to nutrition and food choices, then they must fully understand and be able to communicate the realities of such decisions. Physicians must be required to complete, in either graduate or post-graduate studies, courses on food choices as they relate to our health and the health of our planet. Similarly, dieticians must reread, fully understand, and be able to fully communicate portions of their own position statement on vegetarian diets and the health benefits to their patients. They must do the same with the ill effects that animal products have on our planet—anything less is malpractice. Blood-letting was a common practice for physicians in medieval times, but they would certainly lose their license if they performed that practice today. It is time to move forward with guidance with regard to unhealthy food practices as well.

Government involvement with subsidies and assistance programs for the meat, fish, and dairy industries must come to a halt. If financial support is given to any agricultural industry, it should be to all those crops grown in organic fashion for us to consume. Our government should also be involved in supervising the establishment of proper food education in our school systems, from kindergarten through high school. This could begin by removing all milk and dairy advertisements, and replacing the antiquated and misleading Food Pyramid and food guidelines as established by the USDA with some form of a new pyramid that reflects the revisions established by the PCRM. Promotion should be allowed only for those food choices that are factually healthy for us and for our planet.

There is a perception that our natural resources have no monetary value attached so they can easily be used to create short-term gains by logging forests, misuse of water, wasting land, killing marine wildlife, and polluting. At some point in time, just prior to the enactment of laws against consuming meat, we will need to place an economic value on these resources that have been used so heavily but so freely by the livestock and fishing industries. Specifically, an ecotax, or price, as proposed earlier, should be established and attached to each and every resource that is destroyed or used during the animal-for-food production process. That means that every bit of land, water, and resource used; pollution created; fish species caught inadvertently or that has become endangered; and ancient tree that is cut in the rainforest must be accounted for. Most rainforest trees are over one hundred years old and come with endemic species and diverse ecosystems that are lost as well. What should that price be? And with water, what dollar amount should be placed on the billions of gallons used by the livestock industry, especially when it is from glacier water that could be used by humans directly and that is not readily renewable? When the fishing industry clears a seamount or other area of the ocean of marine life that eliminates communities and destroys the complex ecosystems that have been in place for hundreds of years, what should they pay? Although these resources are essentially priceless and irreplaceable in our lifetime, there must be a form of ecotax in the thousands of dollars for their use, in order to encourage more accountable and sustainable ecological practices.

If I intentionally drove my car off the road and through someone's front yard, destroying every plant, animal, and structure they owned, shouldn't I be responsible for at least paying the cost to replace those items lost? In actuality, it should be my duty to completely restore *everything*, alive or otherwise, back to its original condition and to do so immediately. That should be the case with the fishing and livestock industries. It is unacceptable—economically, philosophically, or otherwise—for a business to use any resource at will and to wipe out species of plants or animals.

The United Nations Committee on Livestock, Environment, and Agricultural Development has stated: "Ultimately, reaching a sustainable balance of demand for livestock and the capacity of ecosystems to provide goods and services in the future will require adequate pricing of natural resources." An ecotax should be levied upon all those that take or destroy our planet's resources, and in turn, this tax should be translated into the final asking price of the product.

In addition to the ecotax, there is reasonable justification to go one step farther by imposing a health-risk tax. This would be the cost a business would have to pay to produce and offer for sale any food item that is associated with a risk for developing chronic disease. Recent estimates place our health care cost burden per year, due to various food choices, at over $100 billion.[163] Because all animal products contribute to the risk of developing the four most prevalent chronic disease conditions, it makes sense that those businesses that offer these food items should help pay for the medical and hospital bills to which it contributes. It is very well documented that the Tarahumara Indians, Seventh-Day Adventists, and other groups of individuals who have eaten only plant-based foods their entire lives do not suffer from adult onset diabetes, triple bypass heart surgery, or colon cancer. Obviously, if all food items made from animal products carried with them this proposed health-risk tax, which is a true reflection of the economic damage created, no one could afford to produce or buy them. This would then help solve the global depletion problem and also vastly improve our health care system, by stimulating less costly health insurance to the consumers who subscribe to consistent, enlightened food choices. It would solve many issues that we, as a civilization, have failed thus far to properly address.

Education must take place in other countries as well. Initial emphasis should be placed on those countries where demand for meat is escalating and also in those countries where indiscriminate destruction of resources occur to support demand for livestock or fish products. A change in demand for animal products is most dramatic in Asia, where consumption of livestock products by humans has increased by 140 percent.[164] Although strong cultural hurdles exist, education of these people to the detrimental effects that animal products have on our planet and to their own health should be the starting point. Equally challenging but necessary is the education of native people in rainforests, where clear-cut logging and erosive, non-sustainable agricultural techniques occur at rampant rates to support the livestock industry and the Western demand for meat. It is mandatory that they understand that the rainforests are invaluable as one of the earth's treasures and, as such, are certainly worth much more economically to them, long-term, by keeping them intact and protected. Global and local education for rainforest economics must be established. It must be understood, for instance, that rainforest land that is converted to cattle operations yields $60 per acre to the landowner for harvested timber, and then the land is worth $400 per acre.[165] However, this same land can produce more than $2,400 per acre if the naturally growing fruits, flowers, and medicinal plants are harvested in a renewable and sustainable manner.[166] Land used in this fashion would be worth substantially more than $2,400 per acre, if it were also used for ecotourism. This could easily be accomplished by combining preservation, sustainable harvesting, and education with controlled tourism. It is vital for local societies to recognize and understand the greater value of natural ecosystems for retaining biodiversity. This would then serve as the impetus toward adequate policy enactment and ultimate preservation, rather than continued destruction.

Once these things are in place, evolution to a more enlightened and healthier route is possible. However, the reality is that this process would require an elongated period of time, during which we would witness continued depletion. Because further global depletion of some category could be devastating, legislation—or a combination of the two—would be the more predictable approach.

So, yes, legislation that bars the raising of animals for food and eating of meat will happen, because it is inevitable that resources such as water, land, food, and energy will be depleted from raising livestock and harvesting fish, to the point where our lives will be affected on a daily basis. Water may very well be the first resource to be affected.

As an example, more than half of the available freshwater supply in the United States is used to grow feed for livestock. Because of this and the fact that this water is nonrenewable, water tables in the Midwest and Great Plains are quickly being depleted, while surface water in the western states is running on borrowed time. Shortages are beginning to occur and will become commonplace, especially in western states. Although consumers in all of these states have been forced occasionally to ration water, they have not been told that the reason they are running out of water for showering or drinking is because most of it has been or is being pumped off for livestock—to grow their feed, for them to drink, or in the slaughtering process.

Soon, a city or municipality somewhere will find itself with a water shortage that cannot be blamed solely on drought. Beyond the narrow view of enforcing rationing, policy-makers for that area will be forced to take a closer look at just where all the water is going. This will inevitably reveal that the vast majority of the water is going to livestock, which then will lead to exposing the reality of our food choices and the true impact on our planet.

Laws may be enacted initially that ensure less use of animals for food, as it would be asking too much of our population to eliminate it entirely. Eventually, however, we will come to the conclusion that only food of plant-based origin can be allowed. Even though this is needed right now, it may take years for the proper wheels to be set in motion for accomplishment.

When logically discussing solutions to the global depletion situation, there are always questions posed by those concerned about the effect these solutions would have on various aspects of life, especially with the point that everyone should just stop eating meat, effective today. The three concerns that are the most common are:

1. What happens to all the animals that are currently being raised for meat if we just stop eating them?
2. What would happen to all those people who make their income by doing something with the livestock or fishing industry? It would be devastating economically.
3. Why can't we just produce and eat organic, grass-fed (pastured) livestock, because isn't this method sustainable?

Regarding the first concern—what happens to all the animals?—a phased-in scenario would occur. There is no likelihood whatsoever that everyone in the world would stop eating meat on one given day. Therefore, a steady decrease in demand would result in steadily fewer and fewer animals raised and fish caught or farmed, until we would be at zero production, which would equate to a near zero environmental footprint from a food production standpoint, with the establishment of sustainable systems.

With regard to the second concern—what would happen to all the people who make their income ...?—this is no different than what has happened with changes over time in technology or with our economy forcing industries that have operated for over one hundred years to either adapt by reinventing themselves or terminate. This would not be the first time it has happened, as we have seen this with countless industries. Most recently, microfiche, various filing systems, and the typewriting businesses had to move over with the advent of computers, and the newspaper industry had to close numerous companies that had been in business since the 1800s, due to advancement of media and advertisement mechanisms like the Internet. This is called progress. And it would be called *proper* progress if these newer industries also cleaned up our planet.

The third concern requires a more in-depth response. Let us consider sustainability as it relates to our food choices. What is it? And exactly who is it that determines what human practices are sustainable? Along my many journeys, I have found that most often, those who use the term "sustainable" are those who truly do not know what is sustainable and, more important, what is not. They simply are not aware of all the variables that need to be factored in when determining true sustainability. It's extremely unfortunate that these same individuals or institutions are placed in a position where public opinion is influenced; even policy-making is based on their opinions. Many examples of this are found routinely in the United States and around the world.

One such example involves the policies with regard to whale killing adopted by a few Caribbean islands. In 2008, St. Lucia was one of six islands that voted to lift the 1986 International Whaling Commission (IWC) ban on hunting whales, which essentially allows Japan to use St. Lucia's surrounding waters to kill whales again commercially or for "scientific research" purposes. In exchange for the use of their waters, Japan subsidizes St. Lucia and five other Caribbean island countries by funneling in $100 million to the economies of these island nations.[167] Japan, Norway, and Iceland now kill collectively three thousand whales annually.[168] Most are still slaughtered by using the painful and inhumane penthrite grenade harpoon technique developed in the 1800s, which is an important topic of discussion in and of itself. During my last conversation with officials of St. Lucia, I asked how they could possibly allow hunting in their waters of these very intelligent, sensitive, and social beings, and I encouraged the officials instead to further invest in the growth of whale watching and ecotourism. Chief Fisheries Officer of St. Lucia, Ignatius Jean, responded with hostility, stating, "We allow harvesting of whales in a sustainable manner ... and both industries [whale killing and whale watching] can coexist in our waters." This is an interesting statement from the one person who influences the policies made to allow the taking of another life off the shores of the island, as no one actually knows the real population numbers of sperm whales.[169] Nor does anyone know the sperm whales' social, feeding, breeding and migratory traditions that have been established for thousands of years.[170] We will not realize that this hunting practice is *not* sustainable until the whales are all gone, as has happened with many other whales and other species. Additionally, how sensible is it to think that a highly intelligent and acutely sensitive creature like the sperm whale would feel comfortable and cooperative with whale-watching boats nearby on any given afternoon, when that same morning its entire pod was attacked and its mate was viciously killed right next to him by a Japanese whale-hunting boat. The reason many whales have become extinct and others are now endangered is because those individuals who influence public opinion or decisions and policies on sustainable practices actually do not have all the answers and thereby miscalculate. In the case of species such as the whale, it is a double miscalculation that results in not only loss of strict numbers but also in the individual act of allowing humans to take the life of another living, peaceful, and innocent being. I believe the rule to follow is that nature has a balancing equation of its own that we humans are incapable of fully comprehending, and that whenever we get involved in this equation by creating subtractions (of land, animals, or other resources), it most likely will generate an irreversible imbalance somewhere—whether or not we are capable of measuring this imbalance. With whales, as with our inanimate resources, even baseline projections of availability are disputable, let alone all those tangible and intangible variables that our human interferences affect along the way.

As it becomes more apparent that our current method of producing livestock is unhealthy for one reason or another, the attention will be turned invariably toward grass-fed, "organic," or essentially pastured animal production. This is already justified as being the healthy alternative to our current practices, as it is purported to be "fully sustainable." Again, who is in the supervisory position to proclaim that this would be sustainable and thereby will misdirect public perception? Why is this endorsed by highly publicized and influential individuals?

This thinking is wrought with many misconceptions that can, for the most part, be grouped as follows:

- That killing and eating any animal is healthier for us than eating plant-based foods, whether or not those animals have eaten pasture
- That somehow transferring the production of animals for food to another mode can be accomplished in a fully sustainable fashion, meaning without the loss of land, water, air quality, or any other resource

Let's look more closely at these misconceptions. The small local farm and grass-fed livestock movement is quickly gaining momentum, in part because of the promotion by various organizations and authors and lecturers, such as Michael Pollan, Mark Bittman, Joel Salatin, and Jonathan Safran Foer. On its surface, this movement appears to be a remedy for much of what they convey as a concern for a healthier diet. After all, modifying our demand for meat to be raised in small farms and on pasture, according to them, accomplishes many things:

- Creates a more "sustainable" way for this type of food to be produced
- Less contribution to pollution
- Provides a "healthier" type of meat
- Breaks down the economic monopoly of our current large agro-businesses in support of the local and small farmer
- Establishes a more humane way for animals to be still used for food

Growing food on a small farm was partially sustainable a hundred years ago—"partially" because at that time, we really did not have a precise method of evaluating the exact effect this style of farming had on individual ecosystems throughout the world. And certainly, eating quantities of animal products was most likely not sustainable to people's health. My grandparents and great-grandparents lived and "sustained" themselves on a small farm. They grew just enough food, some of it animals, to eat off of their own land—as did 37 percent of the U.S. population in 1910, and as did 80 percent of the population in 1870. Today, less than 2 percent garner their income from agriculture, yet the vast majority of our food (84 percent of the total value of food production per year) is now produced by large agro-businesses, which comprise 12 percent of all farming operations in the U.S. (Economic Information Bulletin # EIB-66, 72 pp, July 2010).

On the surface, then, transforming our current agro-business systems to be more local small-farm–oriented is on the right track for many reasons—but not if these systems include raising animals for food. Reduction of the waste (fossil fuel, time, money, etc.) that occurs in transportation, processing, and packaging could be accomplished by becoming more local farm-dependent. Local, small, family farms would also benefit economically, especially if governmental incentives were provided for them. But these incentives should be provided only if they produce food that is the most sustainable for our planet—which would have to be plant-based foods. Raising pastured livestock may seem to be sustainable locally regarding use of resources, but upon closer examination, it is not. And certainly it is not sustainable on a global scale, where more and more people will need to be fed with less land and fewer of our natural resources. Additionally, it is not sustainable for our own health.

Upon closer examination, we can see exactly how grass-fed livestock would affect each of the various areas of global depletion. Land use would simply increase dramatically. We already know how inefficiently we currently use land to raise livestock. And regardless of whether we use mob grazing, juvenile grass growth rotational pasturing, or any other technique to improve land quality while raising grass-fed livestock, it would still require between two and twenty acres of land to support the growth of one cow, depending on which area of the country or world is involved.[171] I found these figures consistent, whether discussing the topic with the more than thirty experts I contacted in agricultural academic institutions or with the many farmers who have been working with grass-fed livestock for the past few decades. Now, on a global scale we will need to multiply the two to twenty acres per cow times the billion that are currently raised in CAFOs (concentrated animal feeding operations, or factory farms), and you will quickly see that there is not enough land on earth—or even *two* earths—to support this. It would require well *beyond* the 30 percent of all the land mass on earth that livestock are using now.

In the United States alone, there are 98 million cattle per year raised for eventual slaughter.[172] Additionally, there are 70 million pigs raised each year for slaughter, and while no objective studies have shown how many acres of land are needed to allow the growth of one pig, a fair assumption would be five to fifteen acres.[173] And since those proponents of grass-fed cows are concerned about keeping their animals "happy," we should indeed include pigs in any of the discussions related to the continual and supposedly sustainable practice of producing only pasture-fed animals. Pigs actually enjoy walks, foraging as they go, and can use all fifteen acres quite easily for feeding, as well as adequate movement. Most would not be aware of this because, unfortunately, most pigs are never allowed the freedom of pasture.

So, let's just do some simple math here. With just the cows and pigs we currently raise to eat each year, placing them all in "fully sustainable" pastured conditions at the appropriate acreage per animal would require 2,520,000,000 acres of land, just in the United States alone (that's 168 million pigs and cows combined, multiplied times an average of fifteen acres per animal required to sustain it). To put this into clear perspective, it's interesting to note that the United States only has 2,260,994,361 total land acres in its entire mass.[174] You can see it is more than absurd, just from a land-use basis only, to presume that somehow eating all pasture-produced meat is even remotely "sustainable." To be consistent, as well as fair to other animals, we can't omit the raising on pasture of the 250 million turkeys, seven million sheep, and eight billion chickens that we consume each year, which would obviously require even more land than has been used in our calculations for cows and pigs.[175] This merely points out the land-mass use requirements. It is not just the quantity, however, but also the quality of land that is heavily impacted by grazing animals. There would be continued and extensive habitat loss, with its subsequent effect on loss of biodiversity and minimizing of oxygen production/carbon dioxide sequestration because of the continued loss of forests. All of this would simply exacerbate our world hunger issues, because land use inefficiencies will continue. In a grass-fed livestock scenario, only a few hundred pounds per acre of animal tissue would be produced, instead of thousands of pounds per acre of plant-based foods, which have more health benefits for us as well.

Whenever any discussion is undertaken regarding sustainable food production operations, let alone "fully sustainable," there must be inclusion of the measurable effects they would have on our water and air quality. The move to pasture-fed cows would, if anything, simply increase the methane production per cow, as it generally requires the rumen bacteria to work longer to digest grass, in order to produce the same energy content found in grain. Some researchers have found, for instance, that greenhouse gas production is 50-60 percent higher in grass-fed beef.[176] We must also remember that when discussing cattle, each animal, when grass-fed, will need to live an additional twelve to thirteen months beyond the ten to twelve months that is considered routine when grain-fed. That means every cow that is producing 50–60 percent more methane will be doing so over twice as long a period. Additionally, every one of the billions of animals raised for us to eat each year, grass-fed or not, also will use oxygen and produce CO_2 as part of their normal respiration process. Knowing livestock's current contribution to our global warming concerns and our need to *reduce* greenhouse gas emissions, continued raising of animals in a grass-fed manner is not healthy or sustainable for our atmosphere.

Those who support continued but pastured livestock use also are completely overlooking the use of water that is still needed by these animals. Remember, the water currently used to support our livestock industry is not sustainable—why would that change with animals raised more on pasture? It would not. There is still the enormous drain on our water supply by the slaughtering and transfer processes requiring as much as another 400 to 500 gallons per cow. And the proponents of this must think that the animals, if free-ranging, will mysteriously not need to drink the same outrageous thousands of gallons of water per animal per year. If anything, the amount would most likely be increased because of the higher activity level of each cow or pig, and they may need to live longer to achieve the appropriate weight gain prior to slaughter. All of this water will need to come from some-where—aquifers or surface water (lakes, ponds, rivers, streams). Whether the animals are grass-fed or not, this is water that could be used directly for human consumption or to much more effi-ciently produce plant-based foods. We have seen that it requires up to 5,000 gallons of water to produce one pound of beef (650–1000 gallons per burger). If this figure is adjusted to reflect pas-ture only that is being fed to the livestock, it will still require 21,000–22,000 gallons over a 24-month period to raise just one cow. That amount of water is the equivalent of a person taking a five-minute shower each and every day for 6.7 years. So, raising grass fed livestock really is not sustainable from a water-usage standpoint.

Then there is the question of how "sustainable" eat-ing grass fed livestock is on our own health. There will be the same overall lack of sustainability or effect on depletion of our own health as we have already seen, despite the myths that have been generated about grass-fed beef. When you hear or read that "grass-fed beef is healthy," it is, in reality, being compared to grain-fed beef, which is not at all healthy to consume. Grass-fed beef has detectable amounts of beta-carotene, slightly less saturated fat, and slightly more vitamin A and E than its grain-fed counterpart. It will, in most cases, have fewer hormones and lower pesticide and herbicide content. All grass fed beef, howev-er, will still contain unwanted high levels of cholesterol, higher than needed levels of saturated fat, and less than one-sixtieth the amount of beta-carotene as most plants, as well as containing cancer-causing agents such as heterocyclic amines and polycyclic aromatic hydrocarbons, protein that has been implicated in numerous disease states and the five leading causes of cancer, minimal vitamins, no fiber, and no phytonutrients or antioxidants to speak of, particularly when comparing it to nearly all plants used as food.

Many of individuals who exhort the virtues of grass-fed livestock make comparisons between our health now and that of our predecessors, who ate food that was not industrialized. So, when looking back to the beginning of the previous century, we might be able to learn just how "sustainable" consuming pasture-fed animals was.

It is true that back in 1900 to 1910, individuals in the United States were eating grass-fed cattle and other types of animals that were not grain-fed. Yearly consumption of meat was 143 pounds per person, but that did not reflect eating poultry or fish because record-keeping for that form of food was vague at best (USDA). Beef accounted for 41 percent of all meat consumed, compared to less than 30 percent as it is today (USDA). It is safe to say that a hundred years ago, individuals in the United States ate their fair share of meat—nearly as much as we eat today—and it was grass-fed. Even though this was long before the commercialization of food, the leading cause of death back then was coronary heart disease, just as it is today—from eating too much saturated fat and cholesterol from all the animal products, and not enough fiber or phytonutrients that can only be found in plants (The Leading Causes of Death 1908–1910, CDC). The only difference was that in 1910, people died at an average age of forty-eight to fifty-one because they did not have all the stents, triple and quadruple bypass surgical procedures, and medications that we have now, which allow us to live another forty years. In fact, the only reason cancers were number eight on the list of most common mortality in 1910, instead of number two as it is today, was because the average age of death was forty-eight for men and fifty-one for women—they simply did not live long enough to develop all the cancers we see today, many of which (including the five most common malignancies) are related to eating animal products. Certainly from a health-risk standpoint, raising and eating grass-fed livestock is not sustainable at all.

Entertainers, policy-makers, or well-known authors may enjoy feeling as if they are authorities on this subject and therefore promote the continued use of raising animals for our food consumption, but the reality is that raising cows, pigs, or any other animals for slaughter and use for our food is *not sustainable at all*. Pasture-fed, organic, grass-fed, or whatever else you would like to call it is simply *not healthy* for anything or anyone involved.

The concept of migrating toward grass-fed livestock is understandable, in that it is much easier for the general public to accept—much easier than the more obvious evolutionary move away from eating animals altogether.

Even though I have presented a comprehensive view of global depletion that is occurring due to our food choices, it is important also to give you an example of what I think *is* sustainable.

Let's create an exercise using a parcel of land as a model, which can then be extrapolated to include all land used for agricultural purposes on our planet, give or take a few variables, depending on the degree of complexity for this exercise.

The best way to do this is to begin by giving each of you a parcel of land that consists of two acres. You can do whatever you would like with it but in the context that it must be used to grow food. Many of you would want to plant pasture grass and then use your two acres to raise livestock, because after all, it's supposed to be very sustainable—and you may still feel the need to eat meat. You could raise one pasture-fed cow on those two acres, and even throw in a few chickens. Your two acres might be enough land in some areas of the country, but in others, you might need to borrow another few acres from your neighbor, just to support that one cow. You will need to supplement the cow with feed and hay over the winter months—where does that come from? Also, remember this cow will need to drink twenty to thirty gallons of water each and every day, and then you'll need to slaughter it, using hundreds of gallons of water in the process. At the end of the two years required for growth, you will be left with about 480 pounds of what some people would call edible muscle tissue—essentially whatever is left over for you that was cut off of the cow's body to consume as food.

Or, alternatively, you could forego the cow/livestock method and use your two acres to grow varieties of plant-based foods; there are many. Take kale, for example. Stated as one of the primary "power foods" on our planet by a number of food and nutrition experts, this plant delivers more nutrients per calorie than any other food. Kale has antioxidants, and among its many micro- and macronutrients, it has a large amount of vitamins K and C and potassium. This plant food has more than sixty times the amount of beta-carotene than grass-fed beef. Kale has at least forty-five different phytonutrients, each of which has been shown to increase immune response, reduce the likelihood of developing cancer, and provide anti-inflammatory properties. Kale also has a perfect ratio of omega-3 fatty acids, and it has fiber, something no grass-fed livestock can provide.

Looking at land use efficiencies and sustainability of resources, one acre will produce, on average, 10,000 pounds of kale in one year, with no (or minimal) water needed during growth, and no water during the "slaughtering" (harvesting) process. Kale will actually continue to grow through extremes of temperatures from minus-5 degrees through 105 degrees F., and after you pick the leaves, it grows new ones. Also, no pathogens, such as H1N1, *E. coli*, salmonella, *Campylobacter*, etc., will ever grow on these plants—as long as there are no livestock farms nearby.

Remember, you have one other acre left over, so my suggestion is to plant a grain, such as quinoa, which is another powerful food that can be grown quite sustainably, yielding 5,000 pounds per acre and providing a gluten-free source of 18 percent protein with a balanced amino acid profile and 14 percent fiber—quite healthy.

So there you have it. With the first method, you have used your two acres of land to create 480 pounds of animal products used as food, but it is a type of protein still implicated in numerous disease states, and along the way you have produced tons of methane and CO_2, and used, at the least, 15,000–20,000 gallons of water.

Or instead, if you used your two acres to grow plants, such as kale and quinoa, you have produced at least ***30,000 pounds*** of food over a two-year period that required no water and caused no greenhouse gas emissions. And the food you ended up with is infinitely healthier for you and for our planet.

To conclude this exercise, I have a novel idea. Grow only plant-based foods, such as kale and quinoa (although there are many other plants) on your two acres, instead of using the land to support one grass-fed cow. Then feed yourself and your family—you could even feed your neighbors' families. But then look at all that leftover kale and quinoa, and take just a moment to box up some of the remaining thousands of pounds of surplus food that you grew, and ship it to all the starving children in Ethiopia. That is my definition of sustainable.

I hope you now have a better understanding of the immense effect your food choices have on the health of our planet. So where do we go from here? What can you do, as an individual, as a consumer? First, you must take yourself, your health, and the health of your planet more seriously. It is not enough to think only about the type of car you need to drive to use less gas, or to change to energy-efficient light bulbs. These are important, but you must look way beyond global warming toward global depletion. Understand and have it clearly imprinted that the choices you make for food to eat today—every meal, every day—had to come from somewhere other than just your grocery store. Ask yourself what resources it took; what was sacrificed to get it to you. Ask yourself about the true cost of that food—what was depleted in its production process?

It is my hope that this book serves as a platform of food-choice enlightenment from which you will keep your awareness antenna up and open for expansion of knowledge, based on accurate and unadulterated information. With a newfound awareness, you can use common sense and make a clear decision to commit to do the right thing regarding food choices—do not go halfway. I like to refer to Tony Horton, creator of the P90X fitness protocol, who continually motivates his audience to "bring it" and "don't just kinda do it." Sure, deciding to not eat meat once a week or only when not with friends is a step in the right direction. But if you take this halfway approach, here's what you are really doing: you are saying, "I read Dr. Oppenlander's book and now realize that eating meat is contributing to global depletion. Therefore, I will do the right thing and eat plant-based foods a couple of times a week to feel good about myself, and then I will continue contributing to global depletion on every other day." Please remember that with *every* bite you take of any animal product, some serious form of global depletion took place, and something had to be sacrificed. There really is no room to go halfway or to "just kinda do it."

Millions of people are influenced by a few who advocate not only eating grass-fed livestock but also that we approach our food choices from other less-than-sustainable concepts. I have a better approach. For instance, instead of "voting with our forks," which we have actually been doing for the past fifty years—and look where it has gotten us—we should actually vote *with our minds* first; then, let *our forks follow*. Also, it is not so wise to eat only foods that your great-grandmother would recognize, because she ate cows, pigs, turkeys, chickens, lamb, and other unhealthy foods obtained from animal parts—not such a good idea.

If you go to the farmer's market, choose plant-based foods. Let the local farmers that you support know you want organically produced food and nothing that came from raising animals, because that uses too much land and water, and affects our atmosphere and our health.

Now for the most important modification of what you hear from many sources: the "go meatless on Monday" campaign. Good; that's terrific. Now you will be contributing to global warming, pollution, and global depletion of our planet's resources on only six days of the week instead of seven. You will be contributing to our national health-care cost crisis to the tune of $140 billion per year, instead of $143–160 billion, and you will be reducing your likelihood of contracting any one of the five leading causes of cancer to only 50 percent, instead of a 58 percent higher risk than if you ate all plant-based foods. Also, by only eating meat six days of the week, your risk of succumbing to coronary heart disease, cerebrovascular disease, diabetes, or any of a number of other diseases might be reduced by a very small amount, but it is still significant.

The point is clear: If you do not eat meat on just *one day* per week, it is certainly better than eating it every day—but not much better, and you are not doing nearly enough. Eating animals is a choice, not a physiological mandate; there is no reason to eat them. Therefore, there is no reason to produce them, particularly knowing how detrimental this practice is to our health and the health of our planet—as well as knowing the benefits of plant-based foods. So, there is no reason whatsoever to advocate going "meatless" on Monday. My strong recommendation is that we *do not* eat "only foods our great-grandmother would eat"; that we vote with our minds and with a new awareness; and that we go meatless *every day*.

Regarding food choices, continue to enlighten yourself; open up and enhance your level of consciousness. Break away from those cultural and media marketing constraints. Do the right thing and commit. Be absolutely consistent with following through, and then feel great about what you are doing. Your body, mind, and spirit will be in a better place—and so will our planet.

CHAPTER XI

Not-to-Read Chapter

A closer look at the animals

"The greatness of a nation and its moral progress
can be judged by the way its animals are treated."
—Mahatma Gandhi

OF COURSE, THE TITLE OF THIS

chapter is facetious; this chapter desperately needs to be read. And this is why: although I have presented the impact our current food choices have on global depletion, somewhere in the mix it would not be right to exclude the reality of animal management. Why? Because it is real. Similar to global depletion, the manner in which we treat animals raised for food is "out of sight, out of mind." Please understand: this book is not about animal rights, although that is a very noble concern. It is about truth, so some mention of the way animals raised for food are treated behind the scenes is in order. I do not want to expound heavily on this topic, however, because, frankly, you might not view the entire contents of this book properly otherwise. Animal rights has inappropriately become a stigma in some venues. The vast majority of humans would rather not hear about their food origins, particularly if it involves inhumane treatment, torture, abominable living conditions, or the pain and suffering of living things. It is much easier to simply turn the other way.

All meat and fish items are products that are derived from animals that are very capable of carrying out thought processes and feeling emotion. So, unfortunately, whenever you eat meat, dairy products, fish, or any part of any animal, more likely than not, you are contributing to abuse of another living thing. Regardless of what you choose to call that part you are consuming (bacon, hot dog, ham, pork chop, sausage, burger, steak, etc.), it is still an animal.

Let's propose that you had a choice of purchasing two different shirts: one was made in a sweatshop overseas by children who live in substandard conditions and are forced against their will to make that shirt. The second shirt was made legally by a reputable company that cares for and pays its employees respectably. No abuse was involved. This shirt was also better for you and our environment, as it is free of dyes, made with organic cotton, and is softer and hypoallergenic. It is economically similar in cost, and in some cases, this second shirt is actually less expensive to you and is always less expensive to our environment. Most of us would choose the second shirt. The same circumstances and choices exist with the food you eat. Animal products can be chosen, all of which come from abusive, inhumane, or unnecessary origins, and eating them will be unhealthy for you and our environment. Or you can choose a plant-based food that has no such baggage attached. Your choice.

As consumers, it is time that we have a conscience; it is long overdue. Ask questions, increase your awareness, and become more savvy in the decision-making process, based on what is in the planet's best interest. In the case of food, that choice is in your best interest as well. One of the greatest injustices our culture has created is the imposition of masking the reality of food origins. For instance, when a child asks where an apple comes from, mothers and fathers provide a fairly accurate and—most important—truthful answer: from a tree. Or where does this broccoli come from? It comes from a garden, and it was grown as a plant in the ground. But when the child asks where a hot dog comes from, parents likely do not respond, "Well, honey, it started with that cute pig that was killed, and parts of its back leg and buttocks were cut and ground up. And before that, the pig was made to live in conditions and treated in a horrible way, prodded and shocked, throat slit, and hung upside down to bleed and to be processed in a meat-packing plant, many times when it is still alive and squirming with pain." Of course you would not tell your child that, and yet *it is the truth*. There is no exaggeration or falsification to that statement. If children knew this, they—or anyone of sound mind—would certainly choose not to eat meat. But parents continue to play the game of avoidance, misinforming their children of the "health benefits" of eating protein, and fabricating names of dead animal parts that are easier for them to deal with. Do yourself a favor, and visit a factory farm and meat-packing plant sometime, and learn firsthand the reality of what you are eating. Then, knowing that this process and end result actually makes you much more likely to develop heart disease, diabetes, etc., and that it is unhealthy for our planet, make the commitment to choose plant-based foods. It is the right thing to do.

Once children make the connection between wonderfully sensitive, docile, and devoted pigs and the bacon they eat, these more enlightened children will *unanimously* choose not to continue eating animals—much as they would not eat their dog. The thought would be repulsive, and it would never even cross their minds.

Quite unfortunately, though, children are often redirected to believe that meat really comes from the grocer, that animals are put on earth for us to eat, and that meat is good for them and that they need it to get protein. When they reach an age where parental influence lessens, however, and more questions are raised about life in general, truths work their way back to the surface, especially where once a seed was placed.

Sustainability has become a critical topic, now and for future generations, regarding all aspects of our life, and throughout this book I have provided clarification of just how our food choices are related. As we have seen, the impetus toward grass-fed livestock is growing in momentum. I have also pointed out that this is a natural path for many to take because it is simply too difficult to let go of the false sense of "needing" versus the reality of "wanting" to eat meat. As I have demonstrated earlier, raising animals on pasture for us to eat is *not sustainable*, despite the fact that nearly all current well-known and influential authors and lecturers are stating the contrary. Subsequently, this type of food production places the animals themselves and how we view their lives in an interesting position conceptually—and here's why: Those who raise grass-fed animals and all the consumers who eat such animals have convinced themselves that they are doing the right thing for our planet and their own health. After all, they are now "growing food" in a "sustainable" manner and doing great things for the world, right? So, the animals themselves, their lives, and their deaths become almost a symbol of this act—ultimate food sustainability—instead of recognition of the individual living things that they are. As a symbol, they can live and be killed without our giving any real thought to what they experience. There is now a higher positioning for moral and even ethical justification in the act of killing them. In fact, though, this becomes almost *more* of an oversight than what they currently experience in CAFOs (concentrated animal feeding operations, or factory farms). In other words, the farmer raises an animal in a pastured situation, rather than being confined, allowing it to live, breathe, experience emotions, develop a thought process, and even a dependency (not strictly for food) with various aspects of its life and relationships. This animal feels things, thinks things, learns, and makes decisions about its actions. Surely some animals do this more than others, but make no mistake—*all* animals raised for us to eat experience these things. If they had their choice, I am quite sure none of these grass-fed animals would choose to die. Moreover, none of them would want us to kill them. Yet we do.

Then, there is the slaughtering process. There is an eerie paradox that occurs when a farmer raises animals year after year, even calling them by name, and then somehow justifies leading them to slaughter, or killing them himself, and then eating them. On local family-owned farms, such as the type many of us feel we should support, smaller animals such as chickens can be killed on site. Larger animals, such as cows and pigs, still have to be rounded up, transported, and killed in conventional slaughterhouses. The most common method of killing chickens on these small farms is by jamming the bird's body into a cone with its head protruding, essentially confining it so the farmer is allowed to more easily slice its throat. The goal is to have the bird to bleed until it dies before moving it into the scalding/defeathering process. It turns out, then, that all grass-fed large animals are still slaughtered in quite inhumane and conventional, painful ways, and all grass-fed small animals are slaughtered in inhumane and painful ways but closer to home. Knowing this, as well as the fact that we simply *do not need* to eat animals and that it is not sustainable, one must ask: how can this be considered a step in the right direction?

It is beyond the scope of this chapter to provide the details of what occurs behind the scenes to raise an animal and process it for eating purposes, but a glimpse of this process is appropriate.

PIGS

Pigs have the cognitive abilities to be quite sophisticated—more so than dogs, and more so than three-year-olds, as observed by many people, including Dr. Donald Broom, Cambridge University professor and former scientific advisor to the Council of Europe.[177] Pigs can play video games, have temperature preferences, and display sensitivities, and they are fond of walking distances of one to three miles per day.[178] In the United States alone, there are over 100 million pigs, with nearly 70 million of these living in factory-farm settings.[179] Each pig living in factory farms is forced to live in an eight-square-foot area for most of its life (which is one to two years).[180] Mother pigs live in a seven-by-two-foot gestation crate, which allows them to give birth and nurse their young but which is too small for them to even turn around.[181] Typical slaughterhouses kill one thousand pigs per hour, making humane death impossible.[182] Many are literally boiled alive or bled to death, with squeals of pain "with numbers squealing reaching 100 percent of those killed during the process in many slaughterhouses."[183]

CHICKENS

As with many animals, there are many misconceptions about chickens. They are inquisitive, have intelligence, and can solve problems.[184] In nature, they form social ladders and friendships, recognize each other, and love and care for their young.[185] There are 10 billion chickens raised and killed for meat in the United States every year.[186] Genetic manipulation and growth-promoting drugs are commonly used to produce fast-growing and larger birds. Because of this, they suffer from skeletal difficulties and heart failure because their hearts and lungs cannot keep up with the forced rapid growth rate.[187] Hens are kept in semi-darkness in battery cages, confining seven to eight to a cage, without the ability to move or spread their wings. Their beaks are cut off, and most suffer from broken bones and eventually are slaughtered after two years of this type of life, although they can live for more than ten years with a more natural life.[188] The process of killing these chickens is gruesome. There is a "catching" stage, where up to six thousand birds are caught per hour and crowded, as many as possible, into crates.[189] A recent industry study concluded that the number of broken bones is unacceptably high. These chickens are then hung upside down in shackles, throats slit, and then immersed in scalding water for feather removal. Many are conscious during the entire process.[190]

TURKEYS

Benjamin Franklin thought the turkey should be the national bird of the United States; Mr. Franklin was a very insightful and intelligent individual. He called the turkey "a bird of courage" and had great respect for its agility, beauty, and resourcefulness. A number of studies have found that turkeys display personality, emotion, and thought process, and can even show a preference for different kinds of music.[191, 192, 193] Turkeys are very gentle birds that enjoy open spaces, can fly up to 55 mph, run at speeds of up to 25 mph, and live ten or more years.[194]

The turkeys that you eat are some of the more than 270 million raised and killed each year that are genetically manipulated animals who have very brief, painful lives.[195] Most spend six months in factory farms, where thousands are packed into dark sheds with no more than 3.5 square feet of space per bird.[196] Handlers cut off portions of the turkeys' toes and beaks with hot blades to keep the crowded birds from scratching and pecking each other to death. With genetic manipulation and large amount of antibiotics, farmers can grow unnaturally large birds in a short period of time, creating turkeys with little room for internal organs, improper bone development, and a number of physical difficulties.[197] Many usually die from organ failure or have broken legs because their bones are unable to support their disproportional weight. Most can hardly walk two steps without falling over, and millions do not even make it past their first few weeks before they die from heart attacks or "starve out," a stress-induced condition that causes them to stop eating.[198] The ones that make it to the slaughterhouses are hung upside down by their weak and crippled legs before their heads are dragged through an electrified "stunning tank," which renders them immobilized but not killed.[199] They then have their throats slit (some are missed) and are then scalded to death in hot defeathering tanks.[200] Many investigations have found turkeys were punched, kicked, beaten with metal rods, and had skulls crushed at Butterball plants and elsewhere, before they make their way to the slaughterhouse.[201]

Following the miserable life and killing process, turkeys eventually—and I find this most ironic—end up as centerpieces for holidays—Thanksgiving, Christmas, and Easter—that represent peace, kindness, and hope. Not exactly what I would consider good karma.

COWS

Cows are inquisitive, clever, and peaceful animals.[202] They show emotion and prefer to spend their time together or with another peaceful animal or human friend, with whom they form a strong bond.[203] Cows form complex social relationships, very much like dogs, licking their companion (animal or human) with their tongues out of love and comfort. Like all animals, cows form very strong maternal bonds with their children and therefore, on dairy farms and cattle ranches, mother cows can be heard crying out for their calves for days after they are separated. They would live normally for more than ten years. The lives we have forced upon them is much different than we would ever inflict upon our dogs.

Cows that survive their short lives in feedlots, dairy sheds, and veal farms face the abominable trip to their eventual death for us to eat. The trip can be days, all without food or water, in open trailers during freezing temperatures, where body parts become frozen against the rails or in their own defecation and urination.[204] In summer months, many collapse from heat exhaustion, so they arrive at the slaughterhouse weak, scared, or downed. The *Journal of Animal Science* reports that 38 percent of all cows that arrive for slaughter show signs of lameness and crippling.[205] None of the cows want to leave the truck, so they are struck with electric prods or dragged off with chains and forklifts. A former USDA inspector relates that "uncooperative animals are beaten and have prods poked in their faces and up their rectums."[206] They then are forced down a chute and shot in the head with a bolt gun meant to stun them, although the lines move quickly and the workers are poorly trained, so many cows are still fully conscious when their throats are cut and limbs are sawed off.[207] As a report in the *Washington Post* describes it: "Within twelve seconds of entering the chamber, the fallen steer is shackled to a moving chain to be bled and butchered by … workers in a fast-moving production line."[208] Ramon Moreno, who has worked in slaughterhouses for twenty years, explained to the *Washington Post* that his job is to cut the legs off the animals, and he "frequently had to cut the legs off fully conscious cows. They blink, make noises … heads moving and eyes wide open and looking around, but the line is never stopped simply because an animal is alive."[209] Martin Fuentes, another slaughterhouse worker, told the *Post* that slowing down the line to ensure that animals are properly killed is unheard of, and workers who alert officials to abuses at their slaughterhouse are at risk of losing their jobs.[210] The meat industry employs many impoverished immigrants who can never complain about poor conditions or cruelty to animals out of fear of losing their jobs and being deported.

The meat and dairy industry is very large and powerful, and they have a great deal of money and lobbying force. They certainly would not want you to know everything about your food choices because you then would stop eating their food. Parents and school systems are equally at fault for not telling the truth to children about their food choices. Again, it's out of sight, out of mind.

These are the realities of your food choices. With every burger, steak, pot roast, turkey sandwich, fried chicken, rib, barbecue, pork chop, bacon, ham, or whatever you want to call it or however you want to cook it, you are perpetuating the demand, which furthers the business of raising animals and then slaughtering them for you to eat. You can turn your head the other way, but the process continues. It continues at the detriment and ill fortune for the animals, for our health, and for the health of our planet. Until this moment, most of you have been comfortably unaware with regard food responsibility and global depletion. What you decide to eat is killing our planet, but it does not have to be that way—if the right choices are made.

A Final Word

"Nothing will benefit human health and increase
the chances for survival of life on earth as much as
the evolution to a vegetarian diet."
—Albert Einstein

I WANT TO EXPRESS MY SINCERE

gratitude to my readers for allowing me to challenge you with a
new way to look at the food you choose to eat, to challenge many
cultural doctrines, and to challenge you to understand and be responsible for the decisions you make regarding food. Much, much
more important than expressing my own appreciation, however,
is that ... our planet thanks you.

END NOTES

CHAPTER I

1   United Nations Food and Agriculture Organization Report, November 2006.

2   US Emissions Inventory, 2008.

3   US Emissions Inventory, 2008.

CHAPTER II

4   *Epidemiology* May 2007;18:373-382.

5   United Nations Food and Agriculture Organization, 2008.

6   Blue Planet Biomes. Amazon Rainforest. 2003.

7   Margulis, Sergio (2004) (PDF). *Causes of Deforestation of the Brazilian Amazon.* Washington D.C.: The World Bank.

8   Turner, K., Georgiou, S., Clark, R., Brouwer, R. & Burke, J. 2004. Economic Valuation of water resources in agriculture, From the sectoral to a functional perspective of natural resource management. FAO paper reports No. 27, Rome, FAO.

9   Chrispeels, M.J.; Sadava, D.E. 1994. "Farming Systems: Development, Productivity, and Sustainability". pp. 25-57in *Plants, Genes, and Agriculture.* Jones and Bartlett, Boston, MA.

10   Pimentel, D., Pimentel, M. "Sustainability of meat-based and plant-based diets and the environment." American Journal of Clinical Nutrition, Vol. 78, 660S-663S, September 2003.

11   Pimentel, D., Pimentel, M. "Sustainability of meat-based

and plant-based diets and the environment." American Journal of Clinical Nutrition, Vol. 78, 660S-663S, September 2003.

12 WorldWatch Institute, "Fire Up the Grill for a Mouthwatering Red, White, and Green July 4th,", 2 Jul. 2003. World Health Organization, 2008.

13 "Soy Benefits", National Soybean Research Laboratory.

14 Steinfeld, H.; Gerber, P.; Wassenaar, T.; Castel, V.; Rosales, M.; de Haan, C. 2006. U.N. Food and Agriculture Organization. Rome. "Livestock's Long Shadow - Environmental issues and options."

15 Steinfeld, H.; Gerber, P.; Wassenaar, T.; Castel, V.; Rosales, M.; de Haan, C. 2006. U.N. Food and Agriculture Organization. Rome. "Livestock's Long Shadow - Environmental issues and options."

16 World Health Organization, 2008.

17 World Health Organization, 2008.

CHAPTER III

18 Siegenthaler, et al., 2005.

19 WorldWatch Institute, "Fire Up the Grill for a Mouthwatering Red, White, and Green July 4th," 2 Jul. 2003. World Health Organization, 2008.

20 UN Food and Agriculture Organization, 2006.

21 United Nations Report, November 29, 2006.

22 United Nations Report, November 29, 2006.

23 United Nations Report, November 29, 2006.

24 United Nations Report, November 29, 2006.

25 United Nations Report, November 29, 2006.

CHAPTER IV

26   Rainforests, National Academy of Science.

27   Moran, E.F., "Deforestation and Land Use in the Brazilian Amazon," Human Ecology, Vol 21, No. 1, 1993.

28   United Nations Environment Program, 2009.

29   Greenpeace, "KFC Exposed for Trashing the Amazon Rainforest for Buckets of Chicken," 17 May 2006.

30   Greenpeace, "Eating up the Amazon." April 2006.

31   Raintree Nutrition, Inc. 1996.

32   Roberts, R. "Amazon Rainforest." Tropical-Rainforest Animals.com.

33   U.N. Food and Agriculture Organization: Amazonian and Caribbean Culture, "Crops of the Amazon and Orinoco Regions: Their Origin, Decline, and Future."

34   U.N. F.A.O: Amazonian and Caribbean Culture, "Crops of the Amazon and Orinoco Regions: Their Origin, Decline, and Future."

35   Taylor, L. *"The Healing Power of Rainforest Herbs."* Square One Publishers, Inc. 2004.

36   Taylor, L. *"The Healing Power of Rainforest Herbs."* Square One Publishers, Inc. 2004.

CHAPTER V

37   Steinfeld, H.; Gerber, P.; Wassenaar, T.; Castel, V.; Rosales, M.; de Haan, C. 2006. U.N. Food and Agriculture Organization. Rome. "Livestock's Long Shadow - Environmental issues and options."

38   Motavalli, J. "So You' re an Environmentalist ... Why Are You Still Eating Meat?" E- The Environmental

Magazine. January/February 2002.

39 Steinfeld, H.; Gerber, P.; Wassenaar, T.; Castel, V.; Rosales, M.; de Haan, C. 2006. U.N. Food and Agriculture Organization. Rome. "Livestock's Long Shadow - Environmental issues and options."

40 *Food Revolution*, John Robbins, 2001.

41 Fleischner, T. "Ecological Costs of Livestock Grazing in Western North America." Prescott College Environmental Program." 1994.

42 Fleischner, T. "Ecological Costs of Livestock Grazing in Western North America." Prescott College Environmental Program." 1994.

43 United Nations Environment Programme (UNEP), 2004.

44 Millenium Ecosystem Assessment (MEA), 2005a.

45 Wassenaar et al., 2006.

46 Eswaran, Lal and Reich, 2001.

47 Reich, et al 1999.

48 Global Footprint Network, Data and Results, 2008.

49 Kelly, S. "Cattle and Sheep Grazing." Land Use History of North America/Colorado Plateau.

50 World Conservation Union (IUCN), Red List of Threatened Species 2008.

51 Chrispeels, M.J.; Sadava, D.E. 1994. "Farming Systems: Development, Productivity, and Sustainability". pp. 25-57 in *Plants, Genes, and Agriculture*. Jones and Bartlett, Boston, MA Rifkin, J. The Guardian, May 17, 2002.

52 Rifkin, J. The Guardian, May 17, 2002.

53 Rifkin, J. "The World's Problems on a Plate: Meat Production is Making the Rich Ill and the Poor Hungry." Guardian of London. May 17 2002.

54   Rifkin, J. "The World's Problems on a Plate: Meat
      Production is Making the Rich Ill and the Poor Hungry."
      Guardian of London. May 17 2002.

55   Rifkin, J. "The World's Problems on a Plate: Meat
      Production is Making the Rich Ill and the Poor Hungry."
      Guardian of London. May 17 2002.

56   Rifkin, J. "The World's Problems on a Plate: Meat
      Production is Making the Rich Ill and the Poor Hungry."
      Guardian of London. May 17 2002.

CHAPTER VI

57   Haliweil, B. "Grain Harvest Sets Record, but Supplies Still
      Tight." Worldwatch Institute, December 12, 2007.

58   North Carolina State University Swine Extension 2003,
      National Research Council.

59   Physicians Committee for Responsible Medicine, American
      Dietetic Association.

60   Opie J, *Ogallala* (Lincoln:University of Nebraska Press,
      2000) 158.

61   Wikipedia.org.

62   Steinfeld, H.; Gerber, P.; Wassenaar, T.; Castel, V.; Rosales,
      M.; de Haan, C. 2006. U.N. Food and Agriculture
      Organization. Rome. "Livestock's Long Shadow -
      Environmental issues and options."

63   Compka, Krchnak and Thorne, 2002; UNESCO, 2005.

64   Turner et al. 2004.

65   Southern Nevada Water Authority, 2008.

66   David A. *"Geology of the Salton Trough,"* Western
      Washington University, May 28, 2005.

67   Brown, L *"Emerging Water Shortages,"* (New York: W.W. Norton & Company, 2006).

68   Langellier, B *"Accidental Oasis,"* Tucson Citizen. Special Report 6-20-06.

69   Carrier, J *"The Colorado:A river run dry."* National Geographic.

70   USDA, 2008.

71   Fradkin, P. *"A river no more: The Colorado River and the west"* (Berkley: University of California Press, 1994).

72   Opie, J. *Ogallala* 41.

73   LEAD, 2006.

74   Erlich *"The Population Explosion"* (New York: Simon and Schuster, 1990) 29.

75   Opie, J. *Ogallala* 152.

76   Opie, J. *Ogallala* 152.

77   Opie, J. *Ogallala* 3

78   National Study of Chemical Residues in Lake Fish Tissue. study ongoing 2004.

79   Neilemann C., Hain S.jk, Adler J. *"In Dead Water: Merging of Climate Change with Pollution"* United Nations Environment Programme 2008.

80   Roberts and Hirschfield, UNEP 2006.

81   Cheung, W. et al. *Intrinsic vulnerability in the global fish catch*, Marine Ecology-Progress Series (2007) 333:1-12.

82   Morato, T. *Vulnerability of seamount fish to fishing.* Journal of Fish Biology 2006 b, 68:209-221.

83   Johnston and Santillo, 2004.

84   "A Run on the Banks: How 'Factory Farming' Decimated Newfoundland Cod." E/ The Environmental Magazine..

85   Food and Agriculture Organization of the United Nations

(FAO) 2008.

86 Broadway, M., Stull, D. "Meat Processing and Garden City, KS: Boom and Bust." *Journal of Rural Studies*, Volume 22, Issue 1, January 2006, Pages 55-66, Michael J. Broadway and Donald D. Stull .

87 FAO.

88 United Nations, 2004.

89 Safina, C. *"The World's Imperiled Fish,"* Scientific American, Nov 1995.

90 Report of the Secretary-General, United Nations General Assembly, Oceans and the Law of the Sea, 14 July 2006, item 150.

91 Hance, J. "One-third of Global Marine Catch used as Livestock Feed." mongabay.com October 30, 2008.

CHAPTER VII
92 Livestock's Long Shadow, 2006. Henning Steinfield, Pierre Gerber, Tom Wassenaar, Vincent Castel, Mauricio Rosales, Cees de Haan.

93 Livestock's Long Shadow, 2006. Henning Steinfield, Pierre Gerber, Tom Wassenaar, Vincent Castel, Mauricio Rosales, Cees de Haan.

94 Congressional Legislative Information (OCIR). "Legislative Hearings and Testimony: Statement of Michael Cook before the subcommittee on livestock, dairy, and poultry and the subcommittee on forestry, resource conservation, and research of the committee on agriculture U.S. House of Representatives." May 13, 1998.

95 "Rearing Cattle Produces More Greenhouse Gases Than Driving Cars, UN Report Warns," UN News Centre, 29

Nov. 2006.

96   Surf or Turf: A shift from feed to food cultivation could
      reduce nutrient flux to the Gulf of Mexico. Simon D.
      Donner, Woodrow Wilson School of Public and
      International Affairs, Princeton University, 410a
      Robertson Hall, Princeton, NJ 08544, USA. Received 26
      May 2005; received in revised form 10 February 2006;
      accepted 26 April 2006.Worldwatch Institute,
      December 2008.

97   Sherman, K., Aquarone, M.C. and Adams, S. (Editors)
      2009. *Sustaining the World's Large Marine Ecosystems.*
      Gland, Switzerland: IUCN. viii+142p.

98   Worldwatch Institute, December 2008.

99   Weiss, K Salmon *Farms are Factory Farms: Dioxins,*
      Pollution and Environmenal Carelessness L.A. Times,
      December 9, 2002.

100   Weiss.

101   EVS Environmental Consultants. *"Impacts of Freshwater*
      and Marine Aquaculture on the Environment:Knowledge and
      Gaps". June 2000, p 7.

102   Costanzo et.al., Marine Pollution Bulletin, Vol 42 (2)
      (2001), pp 149-156.

103   Weiss.

104   Worldwatch Institute, December 17, 2008.

CHAPTER VIII

105   CDC/National Center for Health Statistics.

106   The British Medical Journal, 295 (1987): 351-3°.

107   The American Diabetic Association, "Position of the
      American Diabetic Association and Dieticians of Canada:

Vegetarian Diets, Journal of the American Dietetic Association 103 (2003) 748-65.

108 Ornish D. "Can Lifestyle changes reverse coronary heart disease?" The Lancet 336 (1990): 624-6.

109 American Cancer Society Guidelines on Nutrition and Physical Activity for Cancer Prevention.

110 Yale University, "Animal-Based Nutrients Linked With Higher Risk of Stomach and Esophageal Cancers," news release, 15 Oct. 2001.

111 "Processed Meat May Cause Pancreatic Cancer," Xinhua News 22 Apr. 2005.

112 World Cancer Research Fund / American Institute for Cancer Research. Food, Nutrition and the Prevention of Cancer: a global perspective. 1997 1759 R St. NW Washington, DC 20009. 178-FNS/F27 p 12.a.

113 Arnold, Wilfred Niels (October 2005). "The China Study". Leonardo (MIT Press) 38 (5): 436.

114 PCRM. Vegetarian Foods: Powerful for Health Fact Sheet. 2005.

115 House, L. "7 Foods So Unsafe Even Farmers Won't Eat Them." Planet Green. January 25, 2010.

116 PCRM. Health Concerns About Dairy Products Fact Sheet. 2007.

117 Sampson HA. Food Allergy. Immunopathogenesis and clinical disorders. JAllergy Clin Immunol. 2004; 113:805-819.

118 Host A. Frequency of cow's milk allergy in childhood. Ann Allery Asthma Immunol. 2002;89 (6 Suppl 1):33-7.

119 Gartner LM, Morton J, Lawrence RA, et al; American Academy of Pediatrics Section on Breastfeeding.

Pediatrics. 2005; 115 (2):496-506.

120 Journal of American Dietetic Association. May 2000.

121 Preventing Cancer, PCRM (numerous references).

122 National Cancer Institute Fact Sheet, 2010.

123 CDC Report on Obesity, 2009.

124 Olshansky, et al. 2005.

125 Olshansky, et al. 2005.

126 Preventive Medicine, Nov. 1996.

127 CDC Data and Statistics.

128 Olashansky, et al. 2005.

129 Drug resistant bacteria on poultry products differ by brand, John Hopkins Public Health News Center, 16 Mar. 2005.

130 Drug Resistant Bacteria found in U.S. Meat, Reuters Medical News, 24,May 2001.

131   Silbergeld, E.K. Arsenic in food. Environmental Health Perspectives 112:A338-9.

132 U.S. Dept. of Health and Human Services and U.S. EPA, "What you need to know about mercury in fish and shellfish", Brochure Mar 2004.

133 USDA, Food Safety and Inspection Service, *Foodborne Illness and Disease,* 27 Sep 2006.

134 USDA 2006.

135 CDC U.S. Obesity Trends 1985-2009.

136 USDA 2006.

137 Wuetrich B., "Chasing the fickle swine flu," Science (2003) 299:1502-1505

138 CDC. Key Facts About Swine Influenza. June 2006.

139 USDA, National Agricultural Statistics Service.

140 Tuckman, J. "Four-year-old could hold key in search for

source of swine flu outbreak". *The Guardian* ADA Position Statement, 749. (2009-04-27).

CHAPTER IX

141 Plainview peanut plant raises questions, Connect Amarillo, February 3, 2009.

142 McCormick, L.W. "Source of Salmonella Contamination in Peanut Butter May Be Found." ConsumerAffairs.com. February 26, 2009.

CHAPTER X

143 ADA Position Statement, Journal of the American Dietetic Association. July 2009.

144 ADA Position Statement, 749.

145 World Cancer Research Fund?AICR. Food, Nutrition, and the Preventionn of Cancer. A Global Perspective. Washington, DC:AICR;1997.

146 Deckelbaum RJ, Fisher EA, Winston M, et al. Summary of a Scientific Conference on Prventive Nutrition: Pediatrics to Geriatrics. *Circutlation.* 1999;100:450-456.

147 SaukkonenT, Virtanen SM, Karppinen M, et al. Significance of cow's milk protein antibodies as risk factor for childhood IDDM: Dibetologia. 1998;72-8.

148 WCRF. "Food, Nutrition, Physical Activity, and the Prevention of Cancer: a Global Perspective." November 2007.

149 WCRF. "Food, Nutrition, Physical Activity, and the Prevention of Cancer: a Global Perspective."

November 2007.

150 PCRM. A New Direction: Food Pyramid Yields to PCRM's Power Plate. Spring 2010, Vol. XVX, Number 2.

151 Kradjian, R.M. MD. "Milk: The Deadly Poison."Encognitive.com 2003

152 USDA Economic Research Service.

153 USDA Agriculture Fact Book 1998.

154 USDA Agriculture Fact Book 1998.

155 Gaul, G., Morgan, D., Cohen, S. "No Drought Required for Federal Drought Aid." Washington Post. July 18, 2006. A01.

156 Singer, R. "Want Fries With that Budget Crisis?" opednews.com. September 18, 2008.

157 Oliver, R. "All about: Global Fishing." CNN. March 29. 2008.

158 The Best Energy Bill Corporations Could Buy: Summary of Industry Giveaways in the 2005 Energy Bill. Public Citizen Inc.

159 The World Trade Organization and Fisheries Subsidies: Using trade rules to reverse overfishing and promote sustainable fishing world wide. Sakai, C., Sumaila. U., Hirshfield, M. 2008.

160 Worldwatch Institute. Fact Sheet- Catch of the Day: Choosing Seafood for Healthier Oceans. November 6. 2006.

161 WWF Global Marine Programme. Managing Fishing Fleets.

162 The Humane Society of the United States. An HSUS Report: The Impact of Industrialized Animal Agriculture on World Hunger.

163 Wellman, N., Friedberg, B. "Causes and consequences of

adult obesity:health, social and economic impacts in the United States." Asia Pacific J Clin Nutr (2002) 11 (Suppl): S705-S709.

164 FAO Corporate Document Repository. "Diet, Nutrition and the Prevention of Chronic Diseases." 2003.

165 Amazon Prosperity.

166 Amazon Prosperity.

167 CBS News. "Green Light for Commercial Whale Hunts." Tokyo. April 26, 2010.

168 Sea Shepherd Conservation Society.

169 Taylor, B.L., Baird, R., Barlow, J., Dawson, S.M., Ford, J., Mead, J.G., Notarbartolo di Sciara, G., Wade, P. & Pitman, R.L. (2008). *Physeter macrocephalus.* In: IUCN 2008. IUCN Red List of Threatened Species.

170 Taylor, B.L., Baird, R., Barlow, J., Dawson, S.M., Ford, J., Mead, J.G., Notarbartolo di Sciara, G., Wade, P. & Pitman, R.L. (2008). *Physeter macrocephalus.* In: IUCN 2008. IUCN Red List of Threatened Species.

171 USDA Economic Research Service.

172 USDA/ NASS "Livestock Slaughter 2000 Summary," March 2001.

173 USDA/ NASS "Livestock Slaughter," Monthly Reports, July 2001.

174 USDA Economic Research Service Data Sets.

175 USDA/ NASS "Livestock Slaughter 2000 Summary," March 2001.

176 Raloff, J. "AAAS: Climate-Friendly Dining...Meats. The Carbon Footprints of Raising Livestock for Food." Science News. February 15, 2009.

177 Broom D, Cambridge Daily News, 29, Mar 2002.

178   Food and Agriculture Organization of the United Nations,
      "Pigmeat, Slaughtered/Production Animals (Head) 2002,"
      10 Jun. 2003.

179   Kaufman M, *"Pig farming, growing concern"* The
      Washington Post 18 Jun, 2001.

180   Kaufman M, *"Pig farming, growing concern"* The
      Washington Post 18 Jun, 2001.

181   Kaufman M, *"Pig farming, growing concern"* The
      Washington Post 18 Jun, 2001.

182   Gay, L. "Faulty Practices Result in Inhumane
      Slaughterhouses," Scripps Howard News Service,
      Feb. 2001.

183   Warrick, Washington Post, April 2001.

184   Grimes, W. "If Chickens Are So Smart, Why Aren't They
      Eating Us?" *New York Times*, 12 Jan. 2003.

185   PETA Media Center Factsheets: Poultry and Eggs:
      Industries That Abuse Chickens.

186   USDA National Agricultural Statistics Service.

187   Agricultural Research, May 2000.

188   Singer, P. (2006). *In Defense of Animals*. Wiley-Blackwell.
      p. 176.

189   PETA Media Center Factsheets: Poultry and Eggs:
      Industries That Abuse Chickens.

190   Mench and Siegel.

191   National Wild Turkey Federation, *"All about wild turkey
      facts,"* Nov. 2003.

192   Gerlin A, *"On sale now, top turkey classics,"* Knight Ridder
      Newspapers, 26 Nov. 2003.

193   Hougham A, *"Turkey, not as dumb as you think,"* The Daily
      Barometer 26 Nov. 2003.

194 National Wild Turkey Federation, "All About Turkeys: Wild Turkey Facts," Nov. 2008.

195 National Agricultural Statistics Service, "Turkeys Raised," U.S. Department of Agriculture, 18 Aug. 2009.

196 John C. Voris et al., Turkey Care Practices (Davis, Calif.: University of California, Davis, 1998).

197 Weiss, R. "Techno Turkeys: The Modern Holiday Bird Is a Marvel of Yankee Ingenuity," The Washington Post 12 Nov. 1997.

198 Karrow, J., Duncan, I. "Starve-Out in Turkey Poults," Farm Animal Welfare Research at the University of Guelph (1998–2000) Dec. 1999.

199 PETA Media Center Factsheets. Turkeys: Factory-Farmed Torture on the Holiday Table.

200 PETA Media Center Factsheets. Turkeys: Factory-Farmed Torture on the Holiday Table

201 PETA Media Center Factsheets. Turkeys: Factory-Farmed Torture on the Holiday Table.

202 Rosamund Young, The Secret Lives of Cows, Farming Books and Videos, Ltd: United Kingdom, 2003, p. 5.

203 Jonathan Leake, "Cows Hold Grudges, Say Scientists," The Australian, 28 Feb. 2005.

204 Warrick, J. "'They Die Piece by Piece'; In Overtaxed Plants, Humane Treatment of Cattle Is Often a Battle Lost," The Washington Post 10 Apr. 2001.

205 USDA Information Resources on the Care and Welfare of Dairy Cattle 1996-2002

206 Eisnitz, G. "*Slaughterhouse: The Shocking Story of Greed, Neglect, and Inhumane Treatment Inside the U.S. Meat Industry.*" Prometheus Books, 1997.

207   Warrick, J. "'They Die Piece by Piece'; In Overtaxed
      Plants, Humane Treatment of Cattle Is Often a Battle
      Lost," The Washington Post 10 Apr. 2001.

208   Warrick, J. "'They Die Piece by Piece'; In Overtaxed
      Plants, Humane Treatment of Cattle Is Often a Battle
      Lost," The Washington Post 10 Apr. 2001.

209   Warrick, J. "'They Die Piece by Piece'; In Overtaxed
      Plants, Humane Treatment of Cattle Is Often a Battle
      Lost," The Washington Post 10 Apr. 2001.

210   Warrick, J. "'They Die Piece by Piece'; In Overtaxed
      Plants, Humane Treatment of Cattle Is Often a Battle
      Lost," The Washington Post 10 Apr. 2001.

Index

CPSIA information can be obtained at www.ICGtesting.com
Printed in the USA
LVOW11s0040160515

438672LV00005B/7/P